UFOs

Is There Anybody Out There?

The UFO Phenomenon; Ancient to Modern

Les Williams

Published by Gayles Publishing

ISBN 978-0-9956447-5-5

Contents

Chapter 1: Introduction

"Minds are like parachutes; they only function properly when open" Thomas Dewar

There can be few topics that divide people's opinions and generate more heated debate and argument than that of UFOs and aliens. The very word UFO has become so embroiled in ridicule and derision, it has almost lost its original meaning. For many people the words alien and UFO trigger an instant adverse reaction that conjures up images of a load of crazy people standing on a hill-top in the middle of the night waiting for little green men to land right in front of them and say, "Take me to your leader."

This is unfortunate, as it trivialises something that warrants serious scientific study. In fact, this has become such a problem for serious UFO researchers that they have invented a new and more scientific terminology to describe these phenomena. UFOs are now more commonly referred to as UAPs (unidentified aerial phenomena) and aliens are now often labelled EBEs (extra-terrestrial biological entities). In recent years there has been a shift towards the use of this new terminology, especially amongst serious UFO researchers, but in this book I will use the terms with which most of us are more familiar, and so will refer to them as UFOs and aliens.

Whilst we must allow for mistakes, misinterpretations and in some cases downright lies, we are still left with a vast amount of photographic, documentary and eyewitness evidence to show that people the world over are seeing strange things in our skies that simply defy explanation. Scientists and governments alike have tried to play down these sightings by offering all kinds of theories, from cloud formations and weather balloons to swamp gas and ball lightening. But although the vast majority of UFO reports can be explained by conventional means, many cannot. So, how do we account for these? Are they really visitors from another world, or even another time or dimension?

Given the sheer size of the known universe, it seems most unlikely that Earth is the only inhabited planet in it. One has only to look up at the sky on a clear night to see that there are millions of stars visible to the naked eye, and devices like the Hubble and Kepler space telescopes have enabled us to see much, much further. The universe is a big place and is almost certainly teeming with life, and the more we learn, the more we have to accept that this must be true. The universe is believed to be the result of the big bang, which occurred around 14 billion years ago, and is expanding at a phenomenal rate, just like ripples in a pond into which a stone has been thrown. It is thought to contain over 170 billion galaxies, just like our own Milky Way, which itself contains more than 400 billion stars. Many astronomers now believe that in our own galaxy alone, there are likely to be many millions of planets capable of sustaining life, so the odds in favour of other life in the universe are quite staggering.

Whilst we must accept that this hypothesis is at this time largely theoretical, it is based on scientific observation and study and the indications are that, if anything, these figures may be grossly underestimated. But if there is life in the universe, the host planet would have to offer the right mix of environmental and ecological conditions for it to evolve and develop. A planet's ability to sustain life is dependent on its entire eco-system, as well as its relationship with other planetary bodies and its proximity to its host star. When looking for such planets, astronomers are searching in what they call the 'Goldilocks zone.' This is an orbital position not too close to its host star (so not too hot) and not too far away (and so, not too cold). The Earth is just such a planet, as we orbit our sun at an ideal distance for organic life to grow and thrive. Using Earth as a blueprint, scientists are now actively searching for Earth-like planets elsewhere in the universe, and advances in technology have greatly increased their chances of success. Recent data from the Kepler space telescope has already revealed a number of sun-like stars that have an Earth-like planet in their Goldilocks zones. And they are finding more all the time.

To be sure, not all life in the universe will be intelligent life. Much of it will be microbial or cellular, just like we see under our microscopes here on Earth. It has taken 3.5 billion years for life on Earth to evolve into the diverse complexity that we know today, and it is likely that life on other planets would have to undergo the same, or similar, evolutionary process. What this means is that there is likely to be an abundance of planetary life out there, from very primitive, organic life forms to the highly advanced alien races of science fiction. Primitive life is liable to be far more abundant than intelligent life, just as it is here on Earth, but that still leaves the potential for a huge number of alien civilisations that may have developed the ability to travel through space and time.

At our current level of knowledge and understanding of the laws of physics, the maximum speed possible, as defined by Albert Einstein, is the speed of light, which is 186,000 miles per second. Even if this were achievable, it would take years, decades or even centuries to travel to other stars and galaxies in the universe using conventional means. The closest star to the Earth is Proxima Centauri at 4.5 light years away, which means that it would take 4.5 years to reach it travelling at the speed of light. Given that our own galaxy is around 100,000 light years across, it is not difficult to appreciate the challenges we would face when contemplating the possibility of interstellar travel. The numbers involved are so vast that travel between stars and galaxies would appear to be an elusive pipedream for us humans, and many argue that it would be equally unachievable for distant alien races as it is for us here on Earth.

But what if there were other, as yet undiscovered ways of traversing these vast distances in space and time? We humans like to pride ourselves on our knowledge of the stars and our understanding of the laws of physics, but we may appear almost primitive when compared to an alien race far more technologically advanced than we are. Considering how far we have come in just a few thousand years, how much more knowledgeable would we be if we looked a thousand, or even just a few hundred, years ahead?

Scientists are already exploring potentially new propulsion systems, like nuclear pulse propulsion or gravity distortion, that could be used to power a spacecraft over vast distances in space, and many claim that this has already been achieved by reverse-engineering crashed alien spacecraft. The fact is that a technologically advanced alien race could easily have achieved warp drive capability, with the potential to travel faster than the speed of light. Moreover, leading physicists have discovered that space and time can be bent, like folding a piece of paper, thus shortening the distance between two points in the universe by bringing them closer together. Another potential method would be to use worm holes, which can distort space and time in a way that could facilitate travel over long distances in space by creating a kind of short cut from one part of the universe to another. To the casual reader, this may appear to be pure fantasy and science fiction, but that is no longer the case. Scientists have long theorised about such concepts, but recent study and experimentation has proved that all of these things are possible. So, if an alien race has mastered this technology, why would they not use it to explore the universe, just as we would if we if we were able to do so?

If alien races are visiting the Earth, why are they here and what do they want? Popular movies and books would have us believe that these alien visitors are benign and friendly, here to help mankind and to save us from war and self-destruction. But reports of alien encounters vary a great deal, with some contactees describing them as being kind and caring, often instilling a feeling of calm and tranquillity in people with whom they interact, whilst others say that they were positively hostile and terrifying, with a total disregard for human life or feelings. We could benefit a great deal from alien contact, as many believe we already have, but at the same time hostile aliens could pose a real threat to the security and future of mankind. We are sending probes and signals out into space in an effort to find and make contact with alien races, but this policy may be fraught with risk. As we reach out to offer the hand of friendship are we, in fact, attracting unwanted and potentially hostile attention?

But it may already be too late. There are many who believe that alien beings are here now, and that they have bases in remote areas around the world, and even on the Moon. They believe that our governments are fully aware of this and are working with these aliens on some hidden agenda, and that this fact will soon be disclosed to the public. The veil of secrecy is slowly beginning to lift as military pilots who were once sworn to secrecy are now encouraged to speak openly about their experiences, and these accounts are being shared freely with the media and the UFO community. Ufologists have always claimed that governments are afraid to reveal the truth for fear of mass panic and social and religious breakdown. But all that is beginning to change and even the Vatican, once a fierce opponent of the idea that there may be alien life in the universe, now accepts that this is extremely likely. There seems to have been a shift in attitude in recent years and governments may now be accepting the fact that people should know the truth and are ready for it, and we may now be close to full disclosure.

So, what is the truth in all of this, and how can we even know? The UFO phenomenon is a complex and wide-ranging topic and it is exceedingly difficult to sort the wheat from the chaff. There have been so many reports, ranging from the curiously credible to the completely outlandish, it is hard to know what to believe. We have, over the years, been bombarded with so much information, misinformation and downright lies on the subject of UFOs that it is now almost impossible to unravel the mystery, but we should not give up trying. People have been seeing things in our skies for millennia, and these sightings have been growing exponentially in recent years. More and more people, in all walks of life, are beginning to accept that UFOs are real and that we are, indeed, being visited by extra-terrestrials. Scientists and other prominent people who were once reluctant to go on record on the subject are now willing to speak openly about what they know. The late Neil Armstrong, the first man on the moon and one of the most tight-lipped members of the NASA community, made the following ambiguous statement at a NASA conference in 1994. "There are great ideas as yet undiscovered, breakthroughs available to those who can remove some of truth's protective layers."

Armstrong made no further comment, but many believe this was a cryptic message referring to UFOs and aliens, and the role they played in the Apollo space program.

Whether you believe in UFOs or not, there can be no denying that there is something happening in our skies that we just do not understand, and the phenomenon is not confined to recent times. Reports are more frequent now because we live in a world of mass media, where anything can be on the internet and shared globally within seconds. But in order to fully understand the UFO phenomenon, we should consider these recent encounters in a more historical, or even religious context. The UFO phenomenon dates back many thousands of years and is not confined to the Earth, as there have been numerous strange sightings on Mars, on the Moon and in space. But if there really are intelligent life-forms on other planets, whether they are visiting Earth or not, that means we are not alone in the universe, and that has to be the most significant discovery in the history of mankind.

So, is anybody out there? And if so, are they already here amongst us?

Chapter 2: In the Beginning (UFOs and the Bible)

The UFO phenomenon is by no means a new one, and many people believe that UFOs and aliens have been visiting Earth for thousands of years. But it was not until the publication in 1967 of Eric von Daniken's controversial book, Chariots of the Gods, that the idea really captured the public imagination. The book was heavily criticised by mainstream science but despite this, and despite claims that von Daniken had plagiarised work by previous authors, it quickly became a New York Times bestseller, selling more than 70 million copies worldwide. In it, von Daniken claims that alien beings have been visiting Earth since ancient times and that they were worshipped as gods by our early ancestors. He then goes on to claim that these beings intervened in human evolution, imparting knowledge and wisdom on the human race and kick-starting the process of civilisation. Many of the theories presented in the book have since been disproved and von Daniken himself admits that he made mistakes. But there is still much that remains unexplained, and there is ample evidence to support his theories.

Von Daniken identified numerous ancient and prehistoric sites where archaeologists have found images of strange animals and mysterious humanoid figures, which he suggests are not just imaginary drawings of mythical creatures but are depictions of living entities that these ancient people actually saw and encountered in real life. He claims that these images show what we, today, would call aliens and that these aliens have been visiting Earth for millennia, and continue to do so to this day.

Ancient cave art in Tassili, Algeri, showing strange, helmeted figure

There are many examples of this in ancient and pre-historic art, and these can be found all over the world. In Valcamonica, Italy, we see cave paintings believed to be more than 8,000 years old that show what appear to be human-like figures in tight-fitting suits and wearing what look like dome shaped helmets, some of them holding strange, wand-like devices. We see similar cave paintings in Tassili, Algeria and more again in Niaux, Southern France, to name but a few.

We cannot know for certain whether the beings depicted in these paintings are real or imagined, but it is interesting to note that although very similar in appearance and design, these images were created by ancient societies separated by distance and time, who had no interaction with, or even knowledge of, one another. This has led many to believe that they are depictions of real events, and that they show alien beings who visited Earth many thousands of years ago and interacted with our ancient ancestors. This controversial hypothesis has come to be known as the Ancient Astronaut Theory, and has grown in popularity over the years, spawning hundreds of books and TV documentaries and aggressively challenging mainstream science.

But if our ancient ancestors really were visited by alien beings, what evidence do we have to prove it? In 1976, author and explorer Robert Temple published a controversial book entitled, The Sirius Mystery, in which he hypothesised just such a theory. The book features the mystery of the Dogon tribe of Mali, West Africa, a people not known to the west until the late 19th century. The Dogon seemed to possess an in-depth knowledge of astronomy and the stars that could not be easily explained, and Temple spent some time living amongst them to learn what he could about how they may have gained this knowledge. He spoke at length with the village elders, who told him that the tribe had been visited in ancient times by men from the star, Sirius, and that these celestial visitors taught them the secrets of the universe.

According to Dogon mythology the dog star, Sirius (so called because it sits within the constellation, Canis) has a darker companion that cannot be seen from Earth. What they are describing here is Sirius B, a companion star to Sirius that is, indeed, invisible to the naked eye. Furthermore, although long thought to exist, Sirius B was not discovered by mainstream science until 1862 and was not photographed until 1970. The Dogon described Sirius B as being very dense and heavy, but it was not until the 1920's that modern astronomers were able to confirm this detail by establishing that Sirius B is a small white dwarf star that is about the size of the Earth but with the density of the sun. And that's not all. The Dogon knew that the moon was dry and dead and that Saturn had rings around it. They also had knowledge of the moons of Jupiter and knew that the planets revolve around the Sun. How could they possibly know these things without the tools available to modern science? There were claims that the Dogon acquired this knowledge from European explorers in the 19th century, but this seems unlikely. In the 1940's the Dogon elders told French anthropologist, Marcel Griaule that they had been in possession of this knowledge for hundreds of years and they showed him artifacts, believed to be more than 400 years old, that accurately depict the Sirius constellation, and the cycle of Sirius and Sirius B has been celebrated in their ritual dancing since at least the 13th century.

Almost all of the world's religions feature gods that came down from the sky, as do many myths and legends. It is generally accepted that the cradle of human civilisation was in ancient Mesopotamia, between the rivers Tigris and the Euphrates, and it is here that we find some of the earliest accounts of gods that may have been inspired by visitors from other worlds. In a controversial series of books written by Zecharia Sitchin, the author claims that according to ancient Sumerian legend, alien beings came down to Earth from the planet Nibiru some 450,000 years ago. These beings were called the Anunnaki and they came here to mine gold and other minerals, which they achieved by creating a race of slaves that later developed into the human race. These tales may seem incredibly far-fetched, but we see similar stories in other ancient cultures, and many African tribes, including the Zulus, believe that people came down from the sky to mine gold in ancient times.

Over time, many of these ancient myths and legends found their way into numerous religious texts and may even have inspired some of the stories that we find in the Bible. The epic of Gilgamesh is an excellent case in point. This ancient text from the area around Mesopotamia pre-dates the Hebrew Bible by some 2,000 years, but many of the stories and characters that appear in it bear a striking similarity to those found in the Bible. For example, the epic of Gilgamesh features events that can also be found in the book of Genesis, including the creation of Adam and Eve in the Garden of Eden and the story of Noah and the great flood. Both texts also describe how God imparted laws to a chosen race, and how these laws were recorded on sacred stone tablets. Furthermore, it is interesting to note that the name of the first man in the Anunnaki legends was Adamo, whilst the Hebrew word for first man was Adam, as portrayed in the book of Genesis. It is hard to attribute these similarities to mere coincidence, so it would appear that these ancient texts may have had a common origin and that the one was able to influence the other in some way. The question then arises that if ancient Sumerian mythology was influenced by extra-terrestrial visitations, could some of the stories found in the Hebrew Bible have their origins in UFO folklore, also?

The Hebrew Bible is full of stories and events that would appear to make no sense and to defy any logical explanation, but scholars are now studying these events from a new perspective. In recent times researchers have found new evidence to show that many of the stories found in the Bible may actually be based on real events, and some of its characters based on real people. This has opened up all kinds of possibilities and has proven to be fertile ground for ufologists, who argue that where the Bible describes lights in the sky, a modern observer might see UFOs, and where our ancestors saw angels, we might see aliens. On closer examination, the Bible relates many accounts of strange lights in the sky and of people being raised up to the Heavens, and many believe this to be evidence of ancient UFO activity.

For example, the book of Enoch (not included in the Bible, but found amongst the Dead Sea scrolls) describes how Enoch was taken by the angels and, 'Walked with God, and was seen no more.'

Moreover, in the book of Kings (2:11) we are told of an incident that took place as Elijah was walking with Elisha on the banks of the Jordan river.
'As they were walking along and talking together, suddenly a chariot of fire, and horses of fire, appeared and separated the two of them, and Elijah went up to Heaven in a whirlwind.'

In this instance, something appeared and was seen by both Elijah and Elisha, and it carried Elijah up into the sky, or into some kind of craft. Whatever it was that appeared before Elijah and Elisha that day, it was something that they had not seen before and that they found difficult to describe. If the craft that they saw really was a UFO, we may be looking here at the earliest account of what we, today, would call an alien abduction.

Staying with the book of Enoch, our protagonist once dreamed that he saw, 'Two men, very tall, such as I have never seen on Earth. Their faces shone like the Sun and their eyes were like burning lamps. They stood at the head of my bed and called my name, and I awoke from my sleep to see clearly these men standing in front of me.'
The text goes to say that these beings on took Enoch on a trip around the heavens.

There are many such tales in the Bible, but when searching for evidence of possible UFO activity, there is no better place to look than in the Old Testament book of Ezikiel. The book of Ezekiel describes a series of visions experienced by the prophet whilst in exile in Babylon, and these visions have become a focal point for ancient astronaut theorists. It is suggested by some that these were not visions at all, but real-life experiences, described as visions because that is the only way they could be understood at the time. Some claim that what Ezekiel is describing are real events that terrified and traumatised him into thinking that he was being visited by God. But again, when viewed from a modern perspective, these so-called visions look very much like descriptions of UFO encounters.

In Ezekiel 8:2-3 the prophet describes how he, 'Looked and saw a figure like that of a man. From his waist down he was like fire and from there up his appearance was as bright as glowing metal. He stretched out a hand and took me by the hair of my head and lifted me up from Earth to Heaven.' Could this also be an account of alien abduction?

Ezikiel 10:9-21 relates another curious event, which describes strange beings descending in wheel-like machines. On this, Ezekiel says, 'I looked, and I saw beside the cherubim four wheels, one beside each of the cherubim. The wheels sparkled like topaz and the workmanship of the wheels was like a gleam of beryl, and all four had the same likeness. Their rims were high and awesome, and all four rims were full of eyes all around.'

It has been suggested that the cherubim mentioned above and in other parts of the text are extra-terrestrials, as are the angels that we see elsewhere in the Bible. Furthermore, the wheels described here sound remarkably like the classic, saucer shaped UFOs that we see today, and the eyes all around could be lights or windows. The concept of the UFO as we know it today was completely unknown to the ancients, so they would have no point of reference by which to describe these strange beings and mysterious craft. Hence, an alien being might be seen as a cherubim, and a disc-shaped alien craft might be described as a wheel.

The author goes on to describe these beings in great detail, saying, 'The cherubim went in whatever direction the head faced, without turning as they went. Their entire bodies, including their backs, their hands and their wings were completely full of eyes... Each of the cherubim had four faces; one was that of a cherub, another the face of a man, the third a face of a lion and the fourth that of an eagle. While I watched, the cherubim spread their wings and rose from the ground and as they went, the wheels went with them.'

Ezikiel's vision

He adds, 'In the midst of the living creatures was the appearance of glowing coals of fire, or torches. Fire moved back and forth between them, it was bright and lightning flashed out of it. The creatures were darting back and forth as quickly as flashes of lightning.'

Early Biblical scholars have generally held the view that stories such as these were invented by the high priests in order to make a point or to convey some kind of religious message. But if that were so, why make them so outlandish and difficult to understand or to believe? Their very strangeness and complexity add weight to the possibility that they are describing real events, retold as religious encounters because that was how they were seen by those who experienced them at the time. Elijah and Ezikiel were revered and well-respected holy men who saw the hand of God in everything, so it stands to reason that they would see these events as some kind of divine intervention.

We see more mysterious events in the story of Moses and here again, when viewed from a modern perspective, these events raise some interesting and intriguing questions. Born in Egypt of Hebrew parents, Moses was supposedly cast adrift in a basket, where he was rescued by Pharaoh's daughter and raised in the Egyptian court. In a fit of anger, he kills an overseer for beating a Hebrew slave and is forced to flee to Midian where, after serving in the army, he becomes a shepherd. Here he is visited by God in the form of a burning bush, who tells him that the Israelites are His chosen people and that he must go back to Egypt with his brother, Aaron, to free them from the oppression of Pharaoh.

After much research and debate it is now widely accepted that the Exodus story is based on real events. Archaeologists have recently found an ancient inscription that tells the story of the expulsion of the Hyksos from Egypt at exactly the time of the Exodus, and they have been able to identify the

Pharaoh of the oppression as Ramesses II. It would appear that the expulsion of the Hyksos and the fleeing of the Israelites from Egypt may be two versions of the same story. The Bible's version of this story is that God appeared to Moses in the form of a burning bush, which is why so many rational people find it hard to believe. However, this is not quite the way it is described in the text, where Moses tells us that, 'The bush was all aflame, but was not consumed by the fire.' Here, it clearly states that although the bush appeared to be ablaze, it was not consumed by the fire, so this could not have been a fire in the true sense. This would suggest that what Moses may have seen was a very bright light, projected through a bush or a shrub, giving the impression that it was burning. There may also have been some danger, as the voice issued a warning, saying, 'Come no nearer, take off your sandals, the place where you are standing is holy ground.'

Similar words of warning appear again in other parts of the text. When the Lord appeared to Moses on Mount Sinai, we hear that, 'Mount Sinai was altogether on a smoke because the Lord descended upon it in fire.'

Moses moves closer to the spot where this happened and is told, 'Take heed of yourself that you go not up to the mount, nor touch the borders of it. Whosoever touches the mount shall be surely put to death. There shall not a hand touch it that he shall surely be stoned and shot through. Be it beast or man, he shall not live.'

And in Exodus 19:21 Moses is told to, 'Go down and warn the people not to break through to see the Lord, lest many of them may perish.'

We are later told that after long periods of looking at the cloud, Moses' face developed a peculiar glow and that his sister, Miriam, developed leprosy. There is a very clear message here that being in close proximity to these lights and other phenomena could be very harmful, or even lethal, to humans. The onlookers are repeatedly told to keep a safe distance and are warned of dire consequences if they do not obey. Could this be due to the risk of radiation poisoning? If these lights and voices were alien craft and extra-terrestrials, could the bright lights recounted in these stories be descriptions of UFO encounters, and could the glow described on the face of Moses be the result of exposure to radiation?

We cannot prove that these events were UFO encounters, any more than we can prove that they were visitations from God, but we should not rule out the possibility. We see similar stories in other ancient and religious texts from around the world that on face value, appear to make no sense. The fact is that people and cultures from around the world have recorded encounters with strange beings and have woven them into their religious thinking, and many of these stories are surprisingly similar in content and in nature. We could attribute these similarities to coincidence, but how likely is it that people separated by distance and time would invent stories and characters with such a striking level of commonality?

The objective here is not to claim that these stories really are based on alien visitations, but merely to suggest that they could be. Religious texts are not, and were never meant to be, historical documents, so we cannot consider them in that light. For potential UFO sightings in any known historical context, we should move away from religious texts to real historical accounts that can be accurately dated. The scholars of antiquity and the Middle Ages often recorded such events in a more scientific and factual way, and their observations can generally be considered more reliable. These people were not motivated by religious fervour, but by observing the world in order to better understand it. Although often second-hand and anecdotal, the events they describe more closely resemble modern UFO accounts and may be the earliest reliable records of UFO encounters.

Chapter 3: UFOs in History

The earliest known written account of a UFO is from ancient Egypt and can be found in the Tulli Papyrus. This document was discovered in 1933 by the director of the Vatican's Egyptian Section, Alberto Tulli, from whom it gets its name. It dates from the reign of the pharaoh, Tuthmosis III and gives an account of a strange event that occurred in 1480 BCE.

'In the year 22, of the third month of winter… it was found that a strange, fiery disc was coming in the sky. It had no head and the breath of its mouth emitted a foul odour. Its body was one rod in length. It had no voice.' The text goes on to say, 'After several days had passed, they became more numerous in the sky than ever. They shone in the sky more than the brightness of the sun.'

Sceptics have argued that what is described here is just a meteorite or a comet, but that seems unlikely. During the 18th Dynasty, when this sighting occurred, the Egyptian civilisation was at the peak of its power, with an extensive knowledge of astronomy and the stars, so they would surely have known a meteor when they saw one. This was clearly something different. Something so strange, that they saw fit to record it in this way.

Whilst some of these early accounts are anonymous, many can be attributed to known historical writers. The ancient scholar, Julius Obsequens once wrote that in 216 BCE, strange ships were seen in the sky, with further sightings the following year. We are also told that in 100 BCE, 'In Tarquinia, towards sunset, a round object like a globe or a round or circular shield took its path across the sky from west to east.'

In 336 BCE we hear of an encounter involving the army of Alexander. Italian historian, Alberto Fenoglio records that Alexander's army encountered strange craft during the siege of Tyre, saying.

'Suddenly, there appeared over the Macedonian camp these flying shields, as they have been called, which flew in triangular formation, led by an exceedingly large one. In all there were five of them and they circled slowly over Tyre, whilst thousands of warriors on both sides stood and watched in astonishment.' He then goes on to say, 'Suddenly, from the largest shield came a lightening flash that struck the walls, and they crumbled.'

In another account, we learn that whilst on campaign in Asia in 329 BCE, Alexander the Great was leading his troops into battle against the Scythians when two shield-like craft swooped down from the sky and flew low and fast between the two armies, stampeding the horses and terrifying the men. They spat fire from their rims and made a sound like thunder as they passed overhead.

In 154 BCE more flying shields were seen by opposing sides in a battle and in 104 BCE people in two Italian towns saw mysterious flying objects in the sky. In CE 150, on the road between Rome and Capua, a large object shaped like a piece of pottery appeared overhead and descended down onto the road. On its top it had many colours and it shot out fiery rays of light. The object landed and a figure emerged, which was described as looking like a maiden, dressed all in white. And Roman statesman, Pliny the Elder once wrote that, 'In the consulship of Lucius Valerius and Gaius Marius a burning shield appeared, scattering sparks as it ran across the sky.'

Considering how many ancient documents have been lost and how few have survived, it is quite striking just how many accounts of unexplained phenomena we have from that era. But with the fall of the Roman Empire in the west, Europe entered what has been called a dark age, where writing and scholarly pursuit fell into decline. At this time, it was the monasteries and abbeys of northern Europe that perpetuated the art of writing and it here that we should look for later accounts of unexplained aerial phenomena.

In 664, the Venerable Bede wrote, 'In a monastery at Barking, near the Thames, in the burial ground at night, as the nuns were singing at the graves, they beheld a light sent to them from Heaven like a great sheet, which came upon them. And the light lifted up, moved to the other side of the monastery and withdrew again to the heights.'
Bede is generally considered to be a reliable source and was the foremost scholar and historian of his time, so we have no reason to doubt that what he is describing here was a real event.

Accounts like these appear in a great many medieval manuscripts and with a much greater frequency than in Biblical or ancient texts, but they are often hidden deep within a much larger work. By studying such a text in minute detail for the French magazine, Bres, a Belgian scholar came across the following account, dated from the 9th century.

'It so happened that one day in Lyon, people saw three men and a woman descending from ships in the sky. The whole town assembled around the place, crying that these people were magicians sent by Charlemagne's enemy, Duc de Benevent, to destroy the harvest. The four innocents defended themselves by saying that they, too, were country folk and that they had been kidnapped shortly before by miraculous men who had shown them unheard of wonders and had asked them to tell the world about it.'

The medieval period is a surprisingly rich source of potential UFO accounts and the more we look, the more we find. The eminent scholar and historian, Matthew Paris once wrote that.

'On a day in 1239 a great star, like a torch, appeared over the town of Worcester. It rose in the south and climbed up into the sky, giving out a great light. It was shaped like a head. The front part was sparkling and the back gave off smoke and flashes.

Also in the 13th century, the English cleric, Gervase of Tilbury wrote in his book, Otia Imperialia (also known as the Book of Marvels) how an aerial craft flew over the city of Bristol and got caught on a church steeple, where a man climbed up to try and free it. In the 14th century, Henry Knighton of St Mary's Abbey, Leicester wrote in his chronicles how people saw, 'A fire in the sky, like a burning, revolving wheel or barrel of flame, emitting fire from above.'

Some years later, in 1492, Christopher Columbus reported in his log that, 'At around 10pm, the admiral, standing on the quarterdeck, saw a light in the distance and calling on the mate to look, he saw it, too. It appeared like a bright candle moving up and down, then it vanished, only to reappear immediately. The light then rose out of the water and disappeared into the sky.'

By the 16th century, reports of mysterious aerial phenomena were becoming more frequent and the advent of the printing press around that time accelerated this process even further. Printers flooded the market with books and pamphlets and a media-hungry public just could not get enough of them. In 1557 humanist and scholar, Conrad Lycosthenes, published his Chronicles of Portents and Prophesies in which he describes all kinds of strange events, from the story of Adam and Eve, right up to his own time.

Incredible as it may seem, we even have accounts of these strange craft doing battle in our skies. One of these encounters occurred in Nuremburg in 1561 and is supported by a woodcut illustration depicting the event. The report tells us that on the morning of April 14th, 'A dreadful apparition occurred. There appeared in the middle of the Sun, two blood-red semi-circular arcs, just like the moon in its last quarter.'

The narrative goes on to describe round spheres with a dull, black, ferrous appearance, some in line and some standing alone. These globes began to fight amongst themselves, which they continued to do for over an hour. They then, 'Fell down upon the Earth as if all burned out, and wasted away on the Earth with much smoke.'

Another of these aerial battles occurred in Basel, Switzerland in 1566. A pamphlet published by Samuel Coccius describes strange objects doing battle in the sky over a number of days and reads.

'At dawn on August 7th we saw large black spheres coming and going with a great speed, and precipitating before the Sun, and they chattered as if in fight. Many of them were a fiery red and they soon crumbled, then extinguished.'

Strange Spheres over Basel, 1566

So far, we have considered some of the evidence found in historical or religious texts, but we see ample evidence of this phenomena in the world of art, too. A number of famous paintings from the early renaissance period and beyond, show strange objects in the sky that to the modern observer look very much UFOs. These were mostly religious works, commissioned by the church and under very strict instructions as to their content. Artists of that period were given little or no freedom to express themselves freely in these paintings so we must assume that the commissioning bodies gave their approval to include these objects in their work.

One of the earliest known examples of this is a fresco painted onto the wall of the Decani Monastery in Kosovo. The work dates from around 1350 and shows the crucifixion of Christ. The style of the painting is very typical of the art of the time but in the background can be seen strange, globe-shaped objects, metallic in colour, with lights or flame trailing from behind and with figures shown inside. Moreover, in an altarpiece painted by Italian artist, Carlo Crivelli and dated to 1486, we see a scene of the Annunciation which includes a large, disc-shaped object in the background. The object is breaking through the clouds and is projecting a thin beam of light onto the Virgin Mary. Similarly, an image on the wall of a church in Sighisoara, Romania and dated to around 1523 again shows a picture of the crucifixion and in this case, we see a strange metallic disc hovering over the buildings in the background. Another famous example of possible UFOs in art is in The Baptism of Christ, by Flemish artist, Aert De Gelder. The painting dates from 1710 and in it we see a large, circular craft firing beams of light down onto the figures below and covering the whole area with a rich, warm glow.

By the late 18th century, the idea that there might be life on other planets was gradually gaining momentum and was being debated and argued by scientists and astronomers the world over.

Aert de Gelder's 'Baptism of Christ' circa 1710

In 1793, US exile, Thomas Paine published his book, The Age of Reason, in which he states quite firmly that the world must accept the fact that the Earth was being visited by extra-terrestrials. The book was vehemently contested by many mainstream scientists but sold thousands of copies and in 1802, Paine was invited back to the Unites States by the then president, Thomas Jefferson. Jefferson was a charismatic and popular statesman, well known for his keen intellect and scientific mind and along with other great minds like Benjamin Franklin and John Adams, was a firm believer in what he called, the plurality of worlds.

As printed books became more readily available and with the publication of pamphlets and early broadsheets, reports of strange things in the sky became more frequent. Literacy was on the increase and the new generation of readers could not get enough of these early publications. Sold for as little as a penny and aimed at the masses, these broadsheets were full of local gossip and news of upcoming public executions, but they would often contain a little more. They printed anything they thought may be of interest and it is in these early publications that we often find reports of potential UFOs. A report in the Edinburgh Annual Register for 1767 reported that.

'A large, luminous body appeared in the sky, at first looking like a house of fire, but afterwards took the form of something pyramidal. It rolled forwards with impetuosity until it came to the water of Ericht, up which river it took its direction with great rapidity and disappeared above Blairgowrie.'

In 1878, the January edition of the Denison Daily News reported that Texan farmer, John Marting was out hunting when he saw a dark, circular craft flying overhead. What caught his attention was the speed at which the object was travelling and as it grew closer, it increased in size until its shadow covered the entire field. Five years later, in 1883, astronomer, Jose Bonilla took what is believed to be the earliest photo of a UFO when he observed and photographed a formation of strange objects crossing the Sun.

Another famous case from around this time was reported in the Dallas Morning News in 1897. The story claimed that in August of that year, a cigar-shaped object crashed into a windmill on Judge Proctor's remote Texan ranch. The occupant was killed but, although badly disfigured, witnesses claimed that the creature they found there was not of this world. The story goes that the creature was given a Christian burial at the local Aurora cemetery and the wreckage was thrown into the well, which was promptly sealed up. Years later, in 1935, new owner, Mr Brawley Oates demolished the mill and opened up the well but sealed it up again when he suddenly developed severe arthritis, which he blamed on his time spent working on the site. The well has remained sealed ever since.

As an appendix to this story, in 1973 researchers found the alien's grave clearly marked in Aurora cemetery but when they returned sometime later, the site had been cleared and the grave marker was gone. However, the Texas Historical Commission erected a sign outside the cemetery, outlining the story and marking the incident. This sign was later stolen, and there is still a reward available to anyone who can find it.

As daily newspapers became more widespread, many people claimed that these stories were just hoaxes, invented by journalists to help sell more papers, and that may be so in many cases. But there can be no doubting that the number of reported UFO sightings has been on the increase since the late 19th century, and this trend has continued into modern times.

During the second world war, sightings of strange things in the sky significantly increased as military aircraft flew higher and further than ever before. Furthermore, the advent of radar enabled air traffic controllers to confirm these sightings by checking them against plots on their radar screens. Pilots on both sides in the conflict were seeing strange objects in the sky and were growing ever-more confident in reporting these sightings. For example, German Luftwaffe pilot, Hauptmann Fischer recounted that at 17.30 on 14th March 1942, he was scrambled to investigate a strange blip that appeared on the radar at Banak, Norway. He intercepted the intruder at 10,000 feet and saw what he described as an enormous craft, very long and about 15 feet in diameter, but was unable to get close enough to provide a more detailed description of the object. Shortly afterwards, in November 1942, the tail gunner of an allied aircraft flying over the Bay of Biscay reported that they were being followed by a large, wingless object. This was seen by all of the crew and it followed the aircraft for about 15 minutes, before turning 180 degrees and disappearing.

As the war dragged on, and with far more aircraft in the skies, reports of strange aerial phenomena became more widespread. In November 1942 a Bristol Beaufighter crewed by pilot, Edward Schlueter, observer Donald Meiers and intelligence officer Fred Ringwood was flying on a night sortie north of Strasbourg when they saw eight to ten bright lights off the port wing. They were flying at a very high speed and appeared to be moving under intelligent control and in formation. Schlueter turned towards the lights and they disappeared, before reappearing some distance away. On returning to base Schlueter reported this sighting and taking inspiration from the popular cartoon strip, Smokey Stover, he gave these lights the name, Foo Fighters.

These sightings became more commonplace amongst allied aircrews and were a complete mystery. Former Goon star, Michael Bentine was an RAF intelligence officer during WWII and he recalled how he would often receive reports from airmen that they had encountered this strange phenomenon whilst flying missions over Europe.

'Foo Fighters' shadowing allied bombers during WWII

As the number of reported sightings increased, allied military commanders began to fear that these might be some kind of secret weapon, designed and built by the Germans in an attempt to turn the tide of the war. However, at the end of WWII the allies learned that these lights had been reported by German Luftwaffe pilots, too, and that the Germans found them equally baffling. The reports were investigated at the highest level by both the Allies and the Axis forces, but no explanation could ever be found. At the end of the war, these reports were filed away and would no doubt have been forgotten had it not been for the new wave of UFO sightings, beginning in 1947.

Chapter 4: Modern UFO Encounters (Civilian)

It is generally accepted that the modern UFO era began with the Kenneth Arnold incident, which occurred on June 24th 1947. On that day, civilian pilot, Kenneth Arnold was flying his private plane from Washington State to his destination in Oregon and on hearing reports of a missing aircraft in his vicinity, he decided he would help look for it.

As he approached Mount Ranier at an altitude of 9,200 feet he saw a bright flash, then spotted a formation of nine peculiar looking aircraft flying from north to south, just below his own cruising altitude. The objects were travelling at extremely high speed, occasionally moving around and changing formation. Arnold made a quick calculation of their speed by measuring the time it took them to travel between two points and when he checked his figures on landing, he was able to confirm that they had been travelling in excess of 1,200mph.

Arnold described the objects as metallic in colour, flat and half-moon shaped. The leading edge was curved and the back was convex, like the tail on a Chinese lantern. The curious thing was that they did not fly straight and level but skipped through the air, like a saucer might do if it was thrown across a pond. He told this story to a journalist and when his report appeared in the newspaper the next day, the term, 'Flying Saucer' was born.

At first, the report was thought to be a hoax and many refused to believe Arnold's story. But over time, that changed. His story remained consistent in every detail with each telling and his apparent honesty and integrity convinced more and more people that he may well have been telling the truth. He was even able to produce some excellent drawings of what he had seen. His case was further enhanced when on the same day of Arnold's sighting, Captain E J Smith and his co-pilot on American Airlines flight 105 saw five disc-shaped objects flying in a loose formation in the same area and at very high speed. This was also witnessed by a stewardess on the flight.

Kenneth Arnold holds up a drawing of the UFOs he encountered

The Air Force investigated these sightings and said that what they had seen were geese. But as Kenneth Arnold pointed out, geese are not metallic in colour and they do not fly at 1,200mph! The Kenneth Arnold case was to become a watershed in the study of UFOs. From that point on, sightings would become far more frequent and would increase in complexity in the months and years to follow. The flying saucer had arrived, and it was here to stay.

In the years following the end of WWII, people from all walks of life began to recount their own stories and experiences, and below are just a few of the more famous and best-known accounts.

UFOs Buzz Washington DC, 1952

On the night of July 19th 1952 Edward Nugent, air traffic controller at Washington National Airport spotted seven slow-moving objects on his radar screen. As he joked with his supervisor about flying saucers, two more controllers in the tower spotted a strange bright light hovering in the distance before it suddenly shot away at incredible speed. At the same time over at nearby Andrews Force Base, military radar operators were seeing objects on their screens that moved slowly at first, before flying off at more than 7,000mph. Two F-94 fighter jets were sent up to investigate but each time they closed in on the blips, they quickly moved away and were gone.

The next day, Saturday July 20th, the objects were back and again, two F-94 Starfighter jets were scrambled and a high-speed chase ensued. This time, however, one of the pilots made visual contact with the mysterious lights and gave chase.

"I tried to make contact with the bogies below 1,000 feet but I could not get near them," the pilot later reported. "I was at maximum speed but even then, I was unable to close in. I gave up the chase as I could see that I had no chance of reaching them."

Following this incident newspaper headlines across America were screaming, 'Saucers Swarm Over Capital' and 'Jets chase DC Sky Ghosts.' The public outcry was so great the president Harry Truman ordered an immediate and thorough investigation. However, the Air Force instead called what turned out to be the longest press conference since WWII, at which they claimed that the sightings were due to a weather phenomenon called temperature inversion. This explanation was generally accepted by the mainstream public but was fiercely refuted by the pilots and radar operators involved, who said that they were well aware of the phenomenon known as temperature inversion, but that was definitely not what they had seen.

As far as is known, the in-depth investigation never took place and the Air Force did not even notify its own UFO investigation team that they had dismissed the incident as a natural phenomenon.

Lonnie Zamora, New Mexico 1964

One of the best-known cases in UFO folklore involves a traffic cop, a speeding motorist and a deserted road in New Mexico. On the evening of 24th April 1964, policeman Lonnie Zamora was in pursuit of a speeding car at Socorro, New Mexico when he was distracted by a loud crash and a blueish-orange flame about 1.6km distant. This being a mining area, he initially thought that a dynamite shed may have exploded, so he gave up the chase and turned off the road onto rough ground and drove in the direction of the flame.

He eventually came across what he thought was an upturned car and for a brief moment, he saw two figures that he at first thought must be the occupants of the crashed car but later realised that they were small, like children. It also did not yet occur to him how or why a car would be driving on such rough terrain, so far from the road. The object was in a gulley and when he got out of his car to get a better look, he saw a metallic, egg-shaped object standing on four legs and the two figures were now nowhere to be seen.

As Zamora looked on in disbelief, the object began to make a whining sound and gave off a blueish-orange flame, like the one he had seen earlier, and began kicking up the dust around it. It then took off and moved away slowly in a south-westerly direction. Zamora called the station to report the incident and asked fellow officer, Sam Chavez to join him at the site. When Chavez arrived, the two of them scoured the area and found four indentations in the ground where the legs had stood and that some of the bushes in the area had been charred.

The FBI were informed and they alerted nearby White Sands Missile Base, who sent out Army Captain Richard Holder, along with Special Agent Byrnes to investigate. Holder measured the marks on the ground and took soil samples. He also interviewed Zamora and found him to be believable, but somewhat vague, which is hardly surprising. Shortly afterward, UFO investigators including the great J Allen

Hynek, carried out their own investigation and could offer no explanation for the incident.

Encounter in Livingston Forest 1979

Sixty-one-year-old Scottish forestry worker, Robert Taylor was a practical, no-nonsense man, not given to fantasy or exaggeration and all who knew him described him as completely honest and trustworthy. He went about his daily routine in a simple, matter-of-fact way and although he had heard of UFOs, he had no interest in them whatsoever. But all that was about to change.

On Friday 9th November 1979, Mr Taylor finished his morning break at his home and was returning in his pick-up truck to the forest in Livingston, where he worked. He was checking gates and fences in the area and, having driven as far as he could, he left his vehicle and began walking down the path with his dog, Lara. The dog ran off to explore the undergrowth and as Mr Taylor rounded a bend in the track, he was shocked to see in front of him what he immediately took to be some kind of spacecraft.

The object was dome-shaped and had a band of some sort around the bottom and was either resting on the ground or hovering just above it. He described it as being about twenty feet in diameter and dark grey in colour, with a texture that reminded him of emery paper. It seemed to pulsate and parts of it would change to appear transparent or reflective. There were several stems protruding from it and what looked like propellers on top, although they did not rotate.

Looking on is disbelief he saw two small spheres, about two feet in diameter, approach him from the direction of the craft. Strange legs or struts protruded from these spheres and it was on these that they were rolling towards him. He noticed that as each leg touched the ground, it made a strange sucking or popping noise. There was also a very strong smell, like burning brake linings, which nearly choked him and left a foul taste in his mouth. The rolling spheres stopped, one either side of him and they somehow attached themselves to his trousers, and he could feel himself being pulled towards the craft. Struggling for breath and fighting to free himself, he lost consciousness and fell to the ground.

When he came to, the craft had gone and his dog, Lara was barking anxiously beside him. His head was aching and he still had the foul taste in his mouth and had a raging thirst, and as he tried to get up, he found he was unable to do so. He crawled on his hands and knees for about 100 yards, at which point he found he was now able to stand, so he staggered back to his truck. He tried to radio in but his voice had gone, so he started the truck and went to drive off, but he reversed it into a ditch, where it became stuck. As his house was much closer than the office, he decided to walk home and report the incident from there.

On arriving home, a dirty and distressed Mr Taylor managed to tell his wife what had happened and she wanted to call the doctor, but he dissuaded her from doing so. He went to take a bath and Mrs Taylor called Malcolm Drummond, Head of the Forestry Commission, who said he would be right over to see Mr Taylor, after first calling Dr Gordon Adams to ask him to accompany him. When they arrived at the house, Mr Taylor told them his fantastic story and Dr Adams examined him, finding nothing wrong except for some minor abrasions on his legs. The following day, Mr Taylor took Mr Drummond and Dr Adams to the spot where the encounter had taken place, where they found evidence that would seem to back Mr Taylor's story.

The police were called and the area sealed off and on close examination, investigators found a series of small indentations in the earth, as if something heavy had dug into the ground. When these were measured and recorded, they formed a pattern that exactly matched Mr Roberts' description of the strange spheres and the way they had rolled from the mother ship towards him. The indentations were very precise and uniform and were made at a slight angle, just as a small foot might do if it made contact with the ground in a rolling motion. Moreover, an examination of the trousers Mr Roberts had been wearing revealed tears, as if pulled upwards with some force. The position of these tears fitted Mr Taylor's description of where the object had grabbed him, and the pattern matched holes in his underwear and the marks on his legs. Every piece of evidence seemed to support Mr Taylor's testimony and all of his friends and family, as well as the police and official investigators, were convinced he had been telling the truth.

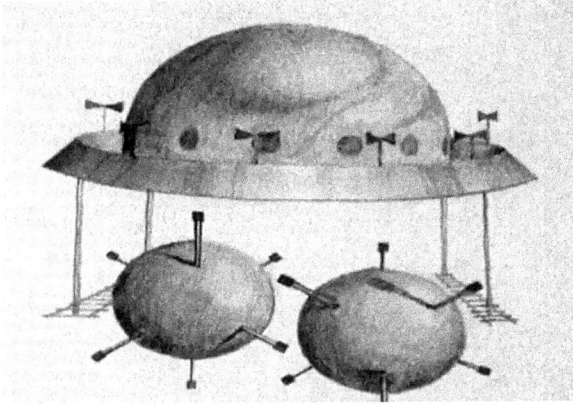

Drawing of the objects that attacked Robert Taylor in Livingston, 1979

Investigators later found indentations in the ground that matched exactly the legs on the smaller spheres

Various explanations have been put forward to try and explain this event, from Mr Roberts having an epileptic fit to ball lightning, but none of them appear to fit. Many agree with Mr Taylor in concluding that what he encountered that day was not of this world. But whatever it was, it was certainly not friendly and submitted Mr Taylor to a terrifying attack and almost succeeded in abducting him.

Cash & Landrum, Huffman, Texas 1980

If you have ever doubted that an encounter with a UFO could be damaging to your health, the following case may change your mind. On the evening of 29th December 1980 Betty Cash, Vickie Landrum and her grandson, Colby were driving along a quiet stretch of road in a dense wood near

Huffman, Texas on their way to their home in Dayton when they saw a light above the trees. They at first thought it to be from an aircraft but when they rounded a bend in the road they saw the light again, this time much closer and brighter. They saw that the light was coming from a large, triangular shaped object, hovering over the trees and emitting a bright flame.

Getting out of the car to investigate, they felt a blast of hot air from the craft that made them break into a sweat, and they felt their skin burning. The object was making a loud beeping sound and the intensity of the light hurt their eyes. Vickie went immediately back to the car to comfort the terrified Colby as the craft flew quickly away, leaving the startled spectators in pain and in a state of shock.

When they arrived at their destination Vickie announced that she had a terrible headache. Colby, too complained of feeling unwell and by midnight, they were both in a poor state. Both had suffered serious burns, had a fever and were vomiting. These symptoms continued for some time, with Vickie's eyesight suffering very badly. Moreover, both Vickie and Colby suffered hair loss which, in Vickie's case, was quite considerable. Her hair did grow back but was then frizzy, where it had once been straight. Colby also suffered nightmares that were so terrifying and severe that Vickie said she thought he might die of fright.

Meanwhile, Betty, too was suffering. She experienced severe headaches which she said were so intense, she thought she was going to die. She had the same stomach problems that had plagued Vickie and Colby and she developed huge water blisters on her head, face and neck, one of which was so large, it obscured her right eye.

Doctors were baffled by these ailments and were unable to determine the cause, although one suspected that it could be some kind of electro-magnetic radiation. Either way, it seems clear that Vickie, Betty and Colby encountered something that day that caused them much suffering and would have a detrimental effect on both their physical and mental health for the rest of their lives.

Todmorden 1980

The following UFO story has it all and involves a policeman, an unsolved murder mystery, a claim of alien abduction and harassment from the men in black. And it all happened in a sleepy little town in northern England.

On June 11th 1980, Yorkshire policeman Alan Godfrey and a colleague were called to investigate a body on top of a coal heap at Todmorden railway station. PC Godfrey had seen many dead bodies before, but this one troubled him a great deal. The body of a man was lying face up on top of a large coal heap and, although it had been raining and the coal was inevitably wet and messy, the man's clothes and hands were clean, which PC Godfrey thought odd, as he had dirtied his own hands climbing up the coal-hill and could see no way this could be avoided. The man's hair had been roughly cropped and there were burn marks around the top if his head and he had an open wound in his neck, which was covered in a mysterious yellow gel. Furthermore, his jacket was buttoned up incorrectly and underneath it he wore no shirt, only a string vest. But the most disturbing feature of the body were the eyes. They were wide open in an expression of sheer terror, and Godfrey could not get this out of his mind.

On close examination, Godfrey concluded that the man had not died on this spot, but had been moved after death, seemingly dropped from above; but how could this be? The body was identified as that of Zigmund Adamski, a Polish mine worker who had gone missing from his home a week before, fully dressed and with a thick mop of dark, wavy hair. A post-mortem examination revealed that Adamski had died from a heart attack and as no explanation could be found for the mysterious event, PC Godfrey confined it to history and continued his normal policing duties.

Some months later, in the early morning of 28th November 1980, Todmorden police had received a number of calls complaining of roaming cattle on the loose and as PC Godfrey was on night duty, he was called to investigate. Following the numerous calls as they came in, he drove around looking for the cattle, but they seemed to elude him every time and he was unable to find them. At around 5am, shortly before the end of his shift, he decided he would drive around and take one more look and as he headed out of town towards open moorland, he could not believe what he saw. There, in front of him, was a craft of some kind hovering silently about 5 feet from the ground. It was diamond shaped, about 20 feet long and 14 feet wide, and had black panels or windows around the top. As it hovered, it spun slowly around anticlockwise and as it did so, the trees and undergrowth were blown in the opposite direction. Godfrey was certain this was a solid object, and he took out his police notebook and made a sketch of it.

There was a sudden and very bright flash of light and the next thing he knew, he was in his police car, some way further down the road and he noticed that his boot was split, but did not know how this had happened. He got out of the car and walked back to where he had seen the object, which was now gone, and he noticed that although it had been raining and the road was wet, there was a dry patch where the object had been hovering. On returning to the station to report the incident, it was noticed that there was a 45-minute time span for which he could not account, and he could not understand why this should be.

Soon after, PC Godfrey was called into his inspector's office, where he was confronted by a man who told him he was from 'The Ministry.' Godfrey took an instant dislike to the man, who had a thick file on his lap containing various papers and documents, including a copy of the sketch Godfrey had made of the object. He was asked all about the encounter and was told not to discuss this, or the mysterious body on the coal heap, with anyone. However, he had already told some of his fellow officers at the station and a colleague alerted the local press, who immediately latched onto the story, linking the UFO sighting with the mysterious case of Zigmund Adamski. It was then just a short step from a local interest story to one of major, international significance. The story was further enhanced by the revelation that Godfrey had, under hypnosis, claimed he had been taken into the spaceship and examined by a group of little creatures and a tall, bearded man.

Godfrey's life became unbearable. He was stalked and followed by the mysterious man from the ministry whom he had met at the station and was constantly hounded by the press. The man even turned up in his local pub, at which point Godfrey told him, in no uncertain terms, to leave him alone. He and his family were harassed from all sides and Godfrey was ridiculed to the point that he lost everything. The incident had cost him his job and his marriage, and he was reduced to sleeping on a friend's sofa whilst drinking a bottle of whisky per day.

Godfrey has said many times that he wishes he had never witnessed these events or been involved in the Adamski mystery, but he maintains that he was and is telling the truth. He says that the incident has ruined his life and that he feels especially sorry for his children, who had to live with not only having a cop for a father, but one who was labelled crazy to boot. Alan Godfrey is a simple, plain-talking Yorkshireman, and not given to fantasy or make-believe. He remains adamant that this thing really happened and, despite all he has suffered and gone through, he would not retract or change his story.

Japan Airlines Flight 1628

On the night of November 17th 1986, seasoned pilot Captain Kenju Terauchi and his two crew members were flying their Japan Airlines 747 cargo plane (JAL 1628) from Reykjavik,

Iceland to Anchorage, Alaska on what should have been a routine flight. At 5.11pm, at 35,000 feet over Alaska, Terauchi noticed two strange objects rise up from below and fly on his left side about 2,000 feet below him. They had glowing nozzles or flames around the edges of what seemed to be rectangular craft, although the objects themselves could not be seen in the darkness. As they drew closer, the lights from the objects lit up the cabin of the aircraft and Captain Terauchi could feel the heat on his face. The objects manoeuvred around in a way that Terauchi described as defying gravity, flying in front of him and assuming a stack formation, at which time Terauchi noticed a third, much larger object following them.

Terauchi called Anchorage Air Traffic Control and was told they could see no other traffic in the vicinity, although the craft were still clearly visible from the windows of the 747. They continued to dance around the aircraft for about 15 minutes before the smaller craft disappeared. Captain Terauchi then noticed a pale band of light that mirrored their altitude, speed and direction and although he could not make out any detail, he determined that there was a large object at about 7.5 miles distance. He again called up Anchorage ATC, who still maintained that they could see nothing on their radar. However, NORAD (North American Aerospace Defence Command) Regional Operations Control Centre (ROCC), directly in his flight path, reported a positive return in their radar systems.

Minutes later, as they passed over the city of Fairbanks, the light from the ground illuminated the object and Terauchi was amazed to see a huge, walnut shaped spacecraft, about the size of two aircraft carriers. Terauchi made a 45-degree turn, descended to 31,000 feet and made a 360 degree turn in the opposite direction and the object remained in formation with his aircraft the whole time. Two other aircraft were vectored in to take a look but they could not see the craft, which had disappeared by the time flight 1628 landed in Anchorage.

When the pilots had filed their reports, FAA official, John Callahan asked the Alaskan regional office to forward the relevant data to their technical centre in Atlantic City, New Jersey, where he and his team played back the radar data and synchronised it with the tapes from the cockpit voice recorder. A day later they briefed Vice Admiral Donald D Engen, who reviewed the data and told them not to talk to anybody, and to prepare a presentation of the data for a group of government officials the next day.

That meeting was attended by representatives of the FBI, CIA and President Reagan's Scientific Study Team, amongst others. Upon completion of the presentation, all present were told that the incident was to be declared top secret and that the meeting "never took place". After viewing the evidence, those present were of the opinion that what the pilots had seen that night was a genuine UFO, most likely of extra-terrestrial origin.

Terauchi's reward for reporting the incident was that he was grounded by Japan Airlines and put into a desk job, although he was returned to flight status several years later. He was, and has always remained, steadfast in his conviction that what he saw that night was not of this world and could only have come from some advanced extra-terrestrial intelligence.

The Giant of Risley, 1987

At 11.30pm on the evening of March 17th 1987, service engineer Ken Edwards was travelling home along the road to Risley in Derbyshire and as he turned off to drive through an old, derelict industrial area, his headlights picked out a strange, lone figure standing on the top of an embankment alongside the old Atomic Energy Authority complex. The figure was over ten feet tall and did not look human. It was wearing a silver-coloured suit with a black helmet that resembled a goldfish bowl and it held its arms out straight, like a sleepwalker. Strangely enough, but the really odd thing was that as the creature began to walk down the grassy slope, it remained at right angles to the ground but did not fall over. Furthermore, it had no knee joints and the legs were articulated from the hip. When the creature reached the road in front of Edwards' car, it turned its head towards him and fired two pencil-like beams through his windscreen, which hit Edwards full in the face.

Edwards claimed that as the beams struck him, his head was suddenly filled with, "Strange thoughts, hundreds of them, all racing through my mind at once. I felt very cold and felt as if two enormous hands were pressing down on me from above. The pressure was tremendous and I was paralysed. I could only move my eyes, the rest of me was rigid."

The creature turned and continued to walk across the road, walking straight through a ten-foot security fence around the atomic energy complex as if it were not there. At this point, Edwards' mind went blank.

The next thing he recalled, Edwards was at home at around half past midnight, but did not know how he got there and could not account for the missing hour between the encounter and arriving home. He told his wife he had seen a silver man and although this might have sounded like an outlandish and bizarre statement, she could see from his shocked and shaken condition that he had clearly witnessed something. Furthermore, his watch had stopped at 11.15. He later found that his fingers were red where he had gripped the steering wheel and the two-way radio in his van ceased to operate. A technician examined the radio and found that the circuit board had been burned out, as if subject to an enormous power-surge through the aerial.

Edwards and his wife, Barbara went to the local police station to report the incident and although the duty sergeant found the story strange and hard to believe, he felt that Edwards was being truthful and sincere. The policeman contacted the atomic energy plant and on recounting the story, Edwards reluctantly agreed to go back to the scene of the encounter with the police officer. When they arrived, they were shocked to be met by more than twenty security guards all carrying batons and when Edwards told them his story, he expected them to laugh or to dismiss it, but not one of them showed any reaction. They carried out a quick search of the area, but refused to go into the trees where Edwards claimed to have seen the creature disappear.

On further investigation, Edwards discovered that there had been a spate of UFO sightings in the area at the time. Police constables Rob Thompson and Roy Kirkpatrick had looked into the case and found that four youths had reported seeing a cigar-shaped craft in the sky on the same day as Edwards' encounter.

But that was not the end of the story for Ken Edwards. Less than a year after the incident, he began to feel ill, complaining of severe fatigue and stomach cramps. He was diagnosed with cancer of the kidneys and although that was treated successfully, the cancer then spread to his throat and within a few years of the incident, Edwards was dead. There is no evidence to link Edwards' cancer with the events of 1987, but we cannot rule out the possibility. Either way, Edwards had an experience that left him traumatised and distressed, and his wife said he was never the same after the incident.

Ilkley Moor Alien Photo, 1987

When, on the morning of 1st December 1987, a serving British Policeman took his camera with him on a walk on the moors hoping to get some good shots of the landscape, he got a little more than he bargained for. As he approached the village of East Morton he saw, through the mist, a small, humanoid figure, green in colour and with an oversized head. As the creature saw the policeman and raised its hand to him, he managed to get off one quick shot with his camera before the creature disappeared. He then saw a saucer-like craft emerge from behind some rocks and fly off into the fog.

As this was the pre-digital age, the policeman went to a nearby town to have his film developed and noticed that, although he took the photo at 8am and the town was just a short walk away, it was now 10am. A walk that should have taken just a few minutes, had apparently taken two hours. He contacted UFO researcher, Jenny Randles to see if she could help him understand what had happened. He had no wish to go public with his story, insisting on anonymity, and he even signed over the copyright of the photo to Peter Hoff of the British UFO Research Association.

The Ilkley Moor alien photo

The policeman, Hoff and Randles proceeded to investigate the affair and returning to the site to compare the photo with nearby rocks, they established that the creature must have been about four feet tall. The officer agreed to be subjected to hypnosis by psychologist, Dr James Singleton, which revealed that he had been taken on board a craft and subjected to a physical examination before being shown images of world destruction.

Oddly enough, although the officer's identity had been known by just a very select few, he claims he was visited at his home by two men in suits, whose ID cards said they were from the MOD (Ministry of Defence). He reluctantly invited them in and they told him that they knew all about the incident on the moor, which surprised him very much. They also knew that he had been working with Randles and Hoff and demanded that he hand over the photo. He told them he no longer had it as he had given it to a friend and to his surprise, they accepted that explanation, not even asking the friend's name. They just stood up and left. The behaviour and mannerisms of the visitors was strange and unusual. They seemed oddly interested in the electric fire in the room and kept asking questions about it.

When Peter Hoff contacted the MOD to enquire about the visitors, they denied all knowledge of the incident or of the strange visitors encountered by the policeman. Moreover, the alien photo was examined by the Kodak laboratory in Hemel Hempstead, who could find no evidence of fakery. Sometime later, US Navy optics expert and UFO researcher,

Bruce Maccabee was asked to examine the photo but he declared that it was too grainy to be tested properly.

Belgian Wave 1989-90

What has now become known as the Belgian wave occurred between November 1989 and March 1990. Throughout this period, mysterious flying objects were observed in the skies over numerous locations in Belgium and were seen, photographed and even filmed by thousands of witnesses. The alarm was raised and military jets scrambled, but the nature and origin of these craft was never determined.

Artist's impression of triangular-shaped UFO as witnessed in the Belgian wave

On 2nd December 1989, numerous large, triangular shaped UFOs were seen in the skies over Liege and military jets were scrambled to intercept them. They were vectored onto their targets by radar and were able to make visual contact with the objects but were unable to get close. Reports continued to flood in and in March 1990, sightings were reported on a massive scale, with an estimated 2,600 UFO sightings in Belgian and German airspace. Most of these sightings were of lights in the sky, manoeuvring around and changing formation, but some were of large, triangular craft with lights along the edges and in the centre. They were seen by many eyewitnesses and were filmed and photographed extensively.

The UFOs involved were huge and were seen to operate at both high and low altitudes, and many witnesses said they had the impression that they were putting on a show, travelling at speeds from a leisurely 30mph to around 1,100mph in the blink of an eye. They would sometimes come very close to the onlookers and occasionally fly silently overhead, and they would remain visible for hours making no effort to conceal themselves.

NATO radar stations at Glons and Semmerzake were again alerted and, once again, F-16 fighter jets were scrambled to investigate. As the F-16s approached the objects, they would fly off at phenomenal speed, only to return seconds later and fly very close to, and in formation with, the jets. Then, after a 70-minute cat and mouse chase, the objects dropped quickly to ground level and disappeared. The pilots of the F-16s said it was as if the objects were playing with them.

The Belgian wave has remained one of the best-documented UFO incidents on record and is backed by an extensive body of eyewitness and photographic evidence. Whatever these craft were, they were far advanced of anything we had at the time and were able to toy with the best of our air defences at will. There were even rumours that one of the UFOs had captured and taken a Belgian fighter jet and its pilot, but this has never been substantiated.

The Falkirk Triangle 1993-97

Lacking the glamour and alure of Roswell, New Mexico, Bonnybridge in Scotland may seem an unlikely candidate for being a UFO hotspot, but that is exactly what happened in the 1990s. The first UFO sighting was reported by businessman, James Walker, who claimed that he was driving along the road between Falkirk and Bonnybridge when he saw a star-shaped object hovering in the road ahead of him and blocking his path. The object remained there for some seconds, before flying off at incredible speed.

Other reports soon followed, with people claiming that a 'howling' UFO had buzzed their cars, and one witness saying they had seen a UFO landing on a golf course. The area soon became a UFO hot spot, with people seeing UFOs of all shapes and sizes and the place soon became a media circus. Local residents were hounded by reporters asking what they had seen and what they thought it might be.

One of the many sightings involved the Slogget family, who were walking on the moors around Bonnybridge one day when the area was suddenly swathed in a bright, white light. They looked up and saw a huge UFO that one of them described as looking like a Tonka toy, emerge from behind some trees. The object was making a loud whirring sound which increased in intensity as it rose higher in the air and with a blinding flash, it descended back behind the trees.

There then followed one of the strangest incidents in the case. One of the UFO enthusiasts visiting the area stopped a passing car to ask if they had seen anything unusual. The occupants said that they had not, and drove off, and the ufologist thought no more of it. However, the following day the police visited him at his home and interrogated him vigorously about the incident, wanting to know if anyone had told him anything or if he had seen anything strange, himself. He was mystified by this, as he had done nothing wrong, but what intrigued him the most was how the visitors knew about the incident and how they knew his address. He had not even given the occupants of the car his name, and there were no witnesses present.

In October 1997, Councillor William Buchman wrote to the Prime Minister, Tony Blair to request a full investigation into the phenomenon, but the request was denied. Despite countless sightings, numerous photographs and the mysterious harrowing of witnesses, the Ministry of Defence concluded that there was nothing to investigate.

Zimbabwe School 1994

In 1994, a group of children at the Ariel School in Ruwa, Zimbabwe witnessed an event that none of them would ever forget. The children aged between 8 and 12 were out enjoying their morning break when they saw a silver craft, with four smaller craft around it, hovering close to the school grounds. They watched in amazement as the craft slowly descended and came to rest on some higher ground close to the school field.

The children ran excitedly to the boundary fence and to their shock and surprise, saw a small creature walking on top on the landed craft, as if working on it. At the same time, another creature emerged from the craft and walked towards the children and gave them what they took to be a friendly wave. The being was described as short, with oval-shaped eyes and was dressed in a tight-fitting black suit, with no apparent seams or fasteners. The creature made direct eye contact with the children and many of the older ones said they could feel some kind of telepathic communication about the damage that humans were doing to the planet. They ran in to tell the teachers what was happening but they did not believe them, and by the time they had been convinced enough to go outside and take a look, the craft and the creatures were gone.

Word soon got around and when some of the parents went to the school to ask about the crazy stories their kids were coming up with, they decided they should investigate further. The children were interviewed individually and they all told exactly the same story. Furthermore, they were asked to draw what they had seen, again with no conferring, and the drawings they produced were remarkably similar. Harvard Professor of Psychiatry, Dr John E Mack risked his career by visiting the school and interviewing the witnesses and was absolutely convinced that they were telling the truth. This did not go down well with Harvard college and they reprimanded Mack for making so bold a statement, but he stuck to his guns and refused to retract what he had said. Moreover, South African journalist, Nicky Carter made a documentary about the incident, and many of the witnesses were interviewed for the film. Carter was impressed by the credibility of the witnesses, saying, "When they were interviewed by Mack, with all his professional skills, it was clear that they were telling the truth. They were so consistent and told their stories with such conviction."

Collage of drawings submitted by pupils at the school

Many years later, American film maker, Randall Nickerson travelled the world to find these children, who had now grown up and moved on, to interview them himself. He found that although they had not kept in touch with one another, their stories had still not changed a bit. They all recounted exactly what they had seen all those years ago, with no embellishment or omissions. However, despite the overwhelming body of evidence and all of the eye-witness testimony, the die-hard sceptics still refused to believe the story. It was clear that in this case they could not use the old favourites like weather balloons or ball lightening to explain the event, so they opted in this case for mass hallucination.

Chicago O'Hare Incident 2006

One might suppose that in the wake of 9/11, if an unidentified object was seen flying over a major civil airport in the United States, the authorities might react in some way, but that was not the case in Chicago on November 7th 2006. At 4.15 in the afternoon of that day, a large, disc-shaped object was seen hovering over gate C-17 of Chicago's busy O'Hare airport and was witnessed, and even photographed, by dozens of people on the ground. The object was described as round, metallic in colour and was visible just below the 1,900ft cloud base for approximately 5 minutes, before rising up into the sky, punching a hole in the clouds as it passed through them. Witnesses included aircraft ground crew, passengers and even pilots standing by for take-off, but despite this, both United Airlines and the FAA refused to investigate the incident, claiming it was a weather phenomenon known as a 'hole punch cloud.'

However, the Chicago Tribune got a hold of the story and filed a freedom of information request, which prompted the FAA to order an internal review of the air traffic control communications tapes for the time of the incident. These recordings revealed in detail the conversations and exchanges between air traffic controllers and other airport personnel, including a call by the airport supervisor to an FAA manager in the control tower in which they discussed the sighting as it happened in real time. These tapes then served to support existing witness testimony and add some weight to the calls for a full and complete investigation, which never took place.

Chapter 5: Modern UFO Encounters (Military)

In the previous chapter we examined several cases of encounters with UFOs, all of which are well-documented and have been extensively researched. But all of these were from civilian sources. For many years now, military personnel from countries around the world have been seeing strange objects in our skies, often interacting and engaging with them in some way. UFOs have been a taboo subject in the military since the earliest days of the phenomenon. Pilots and other military personnel have been reluctant, or even forbidden, to discuss the issue and any reports submitted by them have been vigorously repressed. We can, therefore, assume that the few reports we do have must be the tip of the iceberg and that there must be thousands more that have been withheld from the public.

For many years, military and defence organisations around the world have been silent of the subject of UFOs, but the growing tide of evidence and public pressure are forcing a gradual change in policy. Despite the best efforts of governments and intelligence organisations to deny that UFOs even exist, their once-secret files are beginning to emerge into the public domain. The bigger the secret, the more difficult it is to maintain and in the modern age of mass media and the world-wide-web, it is getting harder all the time. The fact is that governments around the world have been keeping secret files on UFOs for decades, and this has proved to be a real obstacle for serious researchers who just want to establish the truth. But some stories have leaked out, and listed below are some of the most well-known and best documented reports from military sources.

Second Lt George Gorman, 1948

In the fall of 1948 Veteran WWII fighter pilot, George F Gorman was serving in the North Dakota Air National Guard, and it was in that capacity that he engaged in a 25-minute dogfight with a mysterious, unidentified object. The encounter was witnessed by people on the ground and in the air and Gorman later said that he had never seen anything like what he saw that day, and that if anyone else had told him about it, he would have thought they were crazy.

On October 1st that year, Gorman was flying with a group of other aircraft when, on nearing their destination airfield, he decided he would stay airborne a little longer to gain some extra flying time in the clear, cloudless conditions. He flew his P51 Mustang around the skies above Fargo, North Dakota, then radioed the control tower for permission to land. They informed him that the only other aircraft in the vicinity was a small Piper Cub, which he could see about 500 feet below him, and they gave the all-clear for landing. But as he turned in for his final approach, he saw what he believed to be a taillight of another plane passing him on his right, although the tower had no other aircraft on their radar.

Eager to take a closer look, Gorman pulled up and approached to within 1,000 yards of the object, which he described being six to eight inches in diameter, silvery white in colour and blinking on and off. However, as he closed in, the blinking light from the object suddenly became constant and it veered off sharply to the left. Gorman tried to follow and got within about 7,000 feet of the object when it made a sharp turn and headed straight for his aircraft. He pushed the stick forward into a dive and the object flew quickly past, about 500 feet above him. It then turned abruptly and continued on its original course. Gorman once again gave chase and each time he approached, the object veered away. He engaged in this cat and mouse game for 25 minutes before the object shot vertically up into the air at high speed and disappeared. He tried to follow but was unable to keep up. On reporting the incident on the ground, Gorman said that the object made no sound and that he was sure it was under intelligent control and could achieve speeds and manoeuvres impossible for any conventional aircraft.

Gorman was not the only person to see the object that day. It was witnessed by the pilot of the Piper Cub seen by air traffic control earlier, and by the air traffic controllers themselves. Lloyd Jensen and H E Johnson said that they saw both the Piper Cub and the mysterious object in the sky at the same time, and that the object was travelling at a phenomenally high speed.

Thomas Mantell, 1948

Lt Gorman may have been luckier than he realised, as later encounters would often prove highly dangerous and sometimes, even fatal. On a cold January day in 1948, Godman Army Airfield at Fort Knox, Kentucky received a call from the Kentucky Highway Patrol to say that a huge, metallic sphere, about 250ft in diameter, had been seen over the town of Madisonville. At about 1.45pm Sergeant Quintin Blackwell saw the same object from his position in the control tower at Godman Airfield and described it as very white and about a quarter of the size of the full moon, and it appeared to be leaving a trail of green mist. The object remained stationary for an hour, then descended slowly towards the ground before shooting up again at high speed.

At the same time, four F-51 Mustang aircraft of the 165th Fighter Squadron were already in the air and were ordered to investigate. One of the pilots was low on fuel, so he returned to base while the remaining three aircraft made visual contact with the object, which they described as metallic in colour and of tremendous size. They revved up their engines and set off in pursuit, following the object in a very steep climb, but as only one of the F-51s had oxygen on board and his supply was running low, they were ordered to break off the chase and return to base. Mantell, however, continued the chase and as he passed 15,000 feet, he reported that.

"The object is now directly ahead of and above me, moving at roughly my speed. It appears to be a metallic object and is of tremendous size." He added, "I am still climbing. The object is above and ahead of me, still moving at about my speed, or faster. I'm trying to close in for a better look."

The other two aircraft had already given up the chase and their last sighting of Mantell was of his aircraft climbing even higher, levelling off at 30,000 feet, at which point it began to fall out of control in a spiralling motion. When the doomed aircraft reached about 15,000 feet it began to break up and it hit the ground close to a farm, with Mantell's shattered body still inside.

In the meantime, Godman AFB pilot, Lieutenant Clements spotted the UFO and, making sure he had oxygen on board this time, he took off in pursuit. He had seen the object quite clearly from the ground and described it as very large, metallic in colour and reflecting the sunlight like the canopy on an airplane. However, on taking to the air he was unable to locate the object again and despite reaching a very high altitude, there was no sign of the craft anywhere.

The following day, the Louisville Courier carried the headline, 'F-51 and Pilot, Captain Mantell Destroyed Chasing Flying Saucers.' Rumours began to circulate that the fighters had chased a UFO and that the UFO had brought down Mantell's aircraft in an act of aggression, or perhaps in self-defence, but this was fiercely denied by the Air Force. The official report into the accident said that what the pilots had seen was the planet Venus, and that Mantell had blacked out through lack of oxygen at high altitude. However, when it was pointed out that the planet Venus was in a different part of the sky at the time and would have been just a tiny spec of light, if visible at all, the report was changed to say that what they had seen was a weather balloon. This theory, too, was soon dismissed by its critics on the grounds that these were all experienced pilots and unlikely to go off in pursuit of a balloon. Furthermore, no balloon could climb or change direction so quickly, and the witnesses were convinced that the object was under some kind of intelligent control.

The Mantell case was a watershed in UFO reporting and was investigated by the Air Force's Project Sign. They maintained their stance that the object was a balloon but behind the scenes, many said that they knew this was a UFO and that it was in some way responsible for Mantell's crash.

Operation Mainbrace 1952

In September 1952 NATO organised and conducted Operation Mainbrace, the largest peacetime military exercise since WWII. 200 ships, 1,000 planes and 80,000 NATO troops gathered in the North Atlantic to simulate a NATO response to a mock attack by the Soviet Union, a very real prospect at the time. There had been a high number of UFO sightings in the weeks leading up to the operation, prompting one Pentagon official to remark candidly, "Watch out for UFOs when you're out there, guys." The words were spoken in jest, but as it turned out, they proved to be quite prophetic.

The first of the many UFO sightings during the exercise occurred on September 13th when the captain and crew of a Danish destroyer spotted a triangular object moving through the sky at tremendous speed. The craft gave off a blueish glow and was estimated to be travelling at more than 900mph. On September 20th the crew of the US aircraft carrier Franklin D Roosevelt saw a silvery sphere in the sky, which seemed to be following the fleet. Newspaper reporter Wallace Litwin photographed the object and reported the sighting but was told that the object was just a weather balloon. However, on investigating further, Litwin discovered that no weather balloons had been released that day. Later that same day, a thin silvery disc was seen flying low over the fleet, before disappearing into the clouds.

There were many other sightings of strange objects in the sky during Operation Mainbrace, and the theory is that the exercise was being observed by extra-terrestrials, keen to know the extent of our nuclear capability. Following the exercise, the military organisations involved tried to play down reports of UFOs and refused to comment in any way, other than to maintain that all of the sightings could be explained by natural means. However, the sightings were just too numerous and were witnessed by too many people to be brushed under the carpet, so Operation Mainbrace has continued to be studied and debated to this day.

Ota AFB, Portugal, 1982

What follows has become a landmark case in the study of UFOs, as it was witnessed and reported by three experienced military pilots, in broad daylight and in clear skies. On 2nd November 1982 Portuguese Air Force pilot, Julio Guerra was flying his Chipmunk trainer aircraft out of Ota Air Force Base in Portugal when he noticed an object flying slightly below him. Looking closer, he saw that the object was dome-shaped, metallic in colour and had none of the usual aircraft features such as wings or engines. He radioed back to base to ask if they could see any other aircraft in the area and they replied that they had nothing on their radar.

The object approached Guerra's plane and began to circle around him. He tried to manoeuvre his aircraft away from the object but despite his skilful handling of the plane, he could not shake it off. He radioed the tower again to report the sighting, but they just laughed and didn't believe him. "Very well," said Guerra, angrily. "Come and see for yourself."

That challenge was heard by fellow-pilots Carlos Graces and Antonio Gomes, who were also in the air at that time, and they decided to vector onto Guerra's position to investigate. When they arrived at the site a few minutes later they too, saw the craft and so were able to confirm Guerra's report. They described the object as being about ten feet in diameter and looking like a metallic hamburger with a grille around the centre. The top half was shiny and metallic, whilst the bottom was the colour of red wine. It was like nothing they had seen before and was clearly controlled by some intelligent force.

The two aircraft formed up to fly in side-by-side formation, when the object suddenly shot up and flew in between them, affording the pilots an excellent view of the craft. It then began to circle them once more and to move away into the distance before again moving in close, as if it were playing with them. All this time, the pilots were in radio contact with each other and with the ground, but after about 20 minutes of this cat and mouse chase, Guerra decided to take direct action. He turned his plane towards the object and flew directly at it and as he drew closer the object positioned itself slightly above him, then shot off at tremendous speed and was gone.

On landing, the pilots filed reports of the incident but initially heard nothing. They were not even debriefed, which would have been normal practice for military pilots. However, sometime later, the Portuguese Air Force appointed UFO investigator, Jose Sottomayor to investigate the case and he agreed to do so, on the condition that he would have full and complete access to all relevant documents and data, including the pilots' reports. This request bounced up and down the Portuguese Air Force chain of command until it finally reached the desk of Air Force Chief of Staff, General Jose Lemos Ferreira. He studied the files and made the unusual step of releasing everything the Air Force had on the case and asking for a full and open inquiry.

This was almost unheard of in military circles at that time, but Ferreira had good reason for taking such bold action, as he had experienced his own UFO encounter many years earlier. Whilst leading a squadron of F84G Thunderbolt jets on a routine flight in September 1957, Ferreira saw a bright light in the sky ahead of him that seemed to flash and pulsate in reds, blues and greens. Just then, he saw smaller lights emerge from the larger one and begin to circle around it. He had no idea what the object was, but was sure it was no conventional aircraft and believed it could not have been man-made. It was this incident that sparked his interest in the UFO phenomenon, and no doubt influenced his decision to release the files. Ferreira was an intelligent and professional airman and believed that if there are things in our skies that may pose a threat, we need to know what they are.

Following the release of the data a full investigation was carried out but was unable to determine the nature or origin of the mysterious objects. The pilots were interviewed individually and their stories were found to be entirely consistent with one another in every respect. Furthermore, they acted professionally throughout, answering all the investigator's questions in a very calm and precise manner.

There can be no doubt that Guerra, Gomes and Graces encountered something very strange in the air over Portugal that day, but as investigators have been able to offer no explanation for it, the case remains unsolved.

Chapter 6: Abductions and Disappearances

If there is intelligent life in the universe, it stands to reason that there are likely to be many different species, and the evidence we have seen here on Earth would seem to support that hypothesis. When we think of aliens or extra-terrestrials, the image that springs to mind is that of a short grey being with a large head and large black eyes, but there have been many accounts of other alien species besides. In fact, one leading Dutch researcher made a study of alien types reported by witnesses and found a wide diversity of responses, with descriptions ranging from human-like, space-suited figures to grotesque reptilian creatures with scaley skin and claw-like hands. If there is any truth in this diversity of races, it therefore follows that if alien beings are visiting the Earth, it would appear that they are from different worlds and may have very differing agendas. This is evidenced by their behaviour and attitudes of the different alien types towards humans.

Of all the elements of the UFO phenomenon, alien abduction is perhaps the one that sceptics find the most difficult to believe. The phenomenon is widespread and we have countless examples of alien abduction but the stories involved are often so incredible and outlandish, they are very hard to swallow. But that does not mean they are not true. Harvard professor John E Mack, once an ardent sceptic, conducted an exercise in which he interviewed hundreds of alleged alien abductees and his conclusion was that although many of those interviewed were attention-seekers, fantasists or just outright liars, a good many more were not. He was surprisingly impressed by the sincerity and intellect of many of his subjects and he emerged from the exercise feeling there must be some truth in it and that the phenomenon should be studied further.

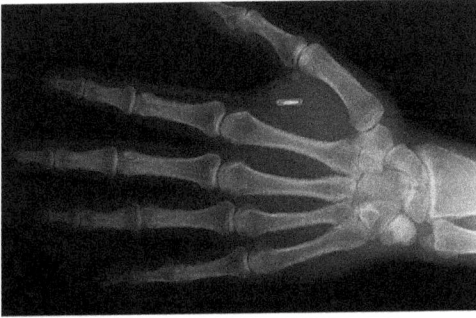

People who claim to have been abducted by aliens will often later find mysterious implants under the skin, as shown in this x-ray of an abductee's hand

It is often argued that following the release of the movie Close Encounters of the Third Kind, reports of grey aliens increased dramatically in number. This is true to some extent, but the grey alien had been with us long before that. The vast majority of alien abduction reports feature the grey alien type, with these creatures often taking people aboard their craft and subjecting them to invasive and often painful examinations. The greys are generally very unpleasant, with no regard for human life or feelings and most people who encounter them suffer severe trauma and distress for many years afterwards. This lack of empathy and compassion has led many to believe that the grey aliens are in fact bio robotic, created by some superior race to do their bidding.

At the other end of the spectrum, we have what are known as the Nordics. These are tall, human-like beings with long blonde hair and bright blue eyes, full of kindness and sensitivity. These beings are reported to be very calm and compassionate and will generally put the abductee at ease through telepathic thought and mind control. Some believe that these are the creators and masters of the greys, as they are often seen together once an abductee has been taken on board the alien spaceship.

Betty and Barney Hill 1961

The earliest, widely reported case of an alien abduction was that of Betty and Barney Hill in 1961. At about 10.30pm on September 19th Betty and Barney Hill, a recently married mixed-race couple from Portsmouth, New Hampshire, were driving home from their honeymoon in Niagara Falls when Betty saw a bright light in the sky. She thought little of it at first but when she realised it appeared to be moving and growing larger, she urged her husband, Barney to stop the car so they could take a closer look. They observed the object through binoculars and saw what they later described as a large craft of some sort passing in front of the moon. Realising this was not a conventional aircraft and fearing for their safety, they got back in the car and drove off, but the object appeared to be following them.

A little further on, the object, which they said was round and appeared to be rotating, descended towards their vehicle and hovered in the road ahead of them, causing the car's electrical system to fail. At this point, they were both overcome with fear and drove off as fast as they could and, before they knew it, they were 35 miles further down the road, with no idea how they got there. When they eventually arrived home, they noticed that their watches had stopped and there were several hours of missing time for which they could not account. Barney's shoes were scuffed on the front and the strap on their binoculars had broken, but neither of them had any recollection of how that had happened. Betty's dress had been torn and was sprinkled with what she described as a pinkish powder. She was also missing the blue earrings she had been wearing that night.

Within days of the encounter, Betty began to experience some vivid and disturbing dreams and when they reported the incident to a group of local ufologists, it was suggested that they should be regressed under hypnosis, to try and unlock their memory of events. Barney was reluctant to do so at first, but Betty persuaded him otherwise and in November 1962, they both agreed to subject themselves to hypnotic regression therapy by Dr Benjamin Simon, a leader in his field.

Over a period of around six months, Dr Simon interviewed Betty and Barney under hypnosis and over time, they began to unlock their sub-conscious memory of what happened to them that night. These sessions were recorded and are readily available on the internet, and they reveal just how terrifying the ordeal was for them. On the tapes, the Hills appear to be extremely distressed and anxious as they describe their terrible ordeal at the hands of the now infamous small, grey aliens. They claim they were subjected to intrusive and very painful physical examination and that the alien beings seemed to have no regard for their pain and their suffering.

To many, their story was just too bizarre and outlandish to believe and even Dr Simons said he did not believe they had really been abducted by aliens, although paradoxically, he also said that he felt sure they were not lying. However, their fear and distress on recounting their experiences under hypnosis seemed all too real, and Dr Simons could not account for this in any way. The Hills' story did change slightly with the passage of time, and this did cast doubt over its authenticity. But there was one feature of the story which is hard to dismiss.

During the regression sessions Betty made the outlandish claim that when they had completed their examinations on her body, the grey aliens told her where they had come from and showed her a map of where they had travelled in the universe. Under hypnosis, Betty was able to reproduce this map in some detail, but astronomers were unable to identify the star system depicted. Then one of them suggested that the map might show a pattern of stars as seen, not from Earth, but from an alien perspective. On further investigation it was found that a likely candidate might be the southern constellation of Data Reticulum, and when this information was fed into a computer at Ohio State University, the map produced was an almost exact match for the one drawn by Betty. Furthermore, Betty's map contained three stars in that constellation that were not discovered by astronomers until 1969, a full six years after Betty had included them on her own diagram.

Moncla and Wilson Disappearance 1953

Air traffic controllers are trained for every eventuality, some of which never happen, but there are occasions when that training is really put to the test. On the night of November 23rd 1953 US Air Defence Command noticed a blip on their radar over Lake Superior, close to Soo Locks, Great Lakes. They were unable to identify the object, so an F-89 Scorpion jet crewed by pilot, Lieutenant Felix Moncla and radar operator, Robert Wilson was scrambled from Kinross Airforce Base to investigate. It was a dark and stormy night but Moncla was an experienced pilot and with Wilson's help was soon able to home in on the object.

The object constantly changed direction as it moved around the sky but with the help of ground radar, the F-89 managed to keep track of it and gave chase, pursuing the object at speeds in excess of 500mph. Ground radar guided them down from 25,000ft to 7,000ft and the jet was finally able to catch up with the object about 70 miles west of Keweenaw Point. Ground radar operators tracked the two plots as they grew closer together and were amazed to see them converge into one and 'lock together.' The radar return for the F-89 had disappeared from the screen and they could no longer make radio contact with the crew. At that point, the plot veered off the screen and it, too, vanished. The US Coast

Guard and Canadian Air Force carried out an extensive search, but no wreckage was ever found.

The official report from the Air Force said that the missing aircraft had been followed on radar until it merged with another object, 70 miles west of Keweenaw Point. The Air Force later retracted that statement, saying that the ground radar was mistaken and that Lieutenant Moncla had suffered an attack of vertigo and crashed into Lake Superior. But if that were the case, there would surely have been some wreckage or debris, however small, to substantiate that claim.

Travis Walton

Former lumberjack, Travis Walton has spent most of his life trying to convince people that an incident that occurred more than forty years ago, really did happen and was very real. On the night of November 5th 1975, Travis and five of his colleagues were travelling home from work in a forest in Arizona when they saw a light shining through the trees. This was in the middle of deer season, so they assumed it must have been hunters stalking their prey with a torch. However, on driving around the next bend they came to a clearing on the side of the road and were amazed at what they found there.

What they had thought was a light shining through the trees was actually a strange, disc-shaped object, hovering a few feet from the ground and about a hundred feet away. It was metallic in colour and was making a deep, rumbling noise, which was felt more than heard. They all stared in amazement and although just as scared as the others, Travis got out of the truck and walked slowly towards the object. His companions were screaming at him to come back, but he continued on and as he did so, the object moved in closer. The object then began to make a very loud noise so Travis dived behind a log for cover, and when he got up to try and run away, he was hit by what he called a strong force and was thrown into the air.

"When it hit me," said Travis, "It was a stunning force. I did not see the blast of energy but the men in the crew gave a statement to the Sheriff's Department and said it looked like a long, blue flame. They were certain it had killed me."
Fearing for their lives and thinking Travis was dead, the rest of the crew sped off to get help and a huge search party was organised, but they found no trace of the missing man.

Meanwhile, Travis claims that he awoke to find himself lying on his back on a raised table inside some sort of spacecraft, surrounded by small creatures that he believed to be aliens.
"It was very blurry," said Travis, "and I had some double vision, but I could see the outline of these forms. At first, I took them to be doctors but as my senses returned, I could see they were not doctors at all. I just flipped out."

Travis panicked and lashed out at the creatures, knocking one of them back into the being standing behind it. He was surprised to find how light the creature was and how easy it had been to push it away. He leapt from the table and, grabbing a tube-like device from a shelf, began waving it at them, which seemed to shock and scare them a great deal. He managed to escape into a different room but was met by a human-like figure dressed in what looked like a spacesuit, and again placed on a table and a mask placed over his face, which rendered him unconscious. The next thing he knew, he was walking down the road, about 15 miles from where he had disappeared. He did not know it at the time, but he had been missing for five days.

The authorities at first suspected that Travis' colleagues may have murdered him and hidden the body but when he turned up, dishevelled but alive, they began to investigate their story more thoroughly. Travis was subjected to intense medical examination and was found to be in good health and was interviewed at length by investigators. Moreover, Travis and his crew were subjected to polygraph tests, which all but one of them passed, the other being inconclusive.

In later years, Travis wrote a book about his experience, which was later made into a film entitled, Fire in the Sky. Both of these were well received but failed to convince the sceptics. There were many who refused to believe Travis' story, and they cited a number of reasons for this. Firstly, Travis Walton was known to have been a UFO enthusiast, having shown a keen interest in the subject for many years and some said that he had made the whole thing up. This view is supported by the fact that just two weeks before the incident, a major TV network screened a documentary on the Betty and Barnie Hill abduction case, which Travis is likely to have watched. Moreover, the team were behind on their contract and were about to incur financial penalties for the delay, so any excuse to have the contract cancelled (which it later was) would have been most welcome.

This is not to say that Travis Walton was lying. He and his colleagues have remained adamant that what they saw and experienced that night was very real, and Travis has spent the rest of his life defending his claims. Like many abductees, he says that he found the whole experience extremely traumatic and wishes it had never happened. Travis says, "From the beginning it was a battle against people trying to explain it away. The locals didn't want to believe it and the sheriff at first treated it as a murder case before later changing his mind, suspecting instead a drug-induced hallucination. I had a whole battery of psychiatric tests and there was nothing wrong in that department. And drugs tests proved negative." He then went on to say, "I took each and every theory the sceptics came up with and just blew them all out of the water with facts."

Travis Walton is a perfectly sane and rational man who just happened to have a very harrowing and traumatic experience that changed him forever. If the story is a hoax, he hides it very well and has been doing so for many years. He maintains that he wishes the event had never happened and that it has ruined his life.

Abduction in Brazil, 1977

Antonio La Ribia did not believe in UFOs and had never given them a second thought, but all of that was about to change in a most traumatic and distressing way. And although he still knows nothing about UFOS, he now knows for certain that they exist and are very real.

Early on the morning of Thursday, September 15th 1977, La Ribia got up out of bed and went to catch the bus to the Oriental Bus Company depot in Paciencia, Brazil, where he was employed as a bus driver. As he walked down the darkened road to the bus stop, he noticed something large hovering silently in a nearby field. He was startled and wanted to run but just then, the whole area was lit up by a bright blue light and three robot-like beings appeared in front of him. They were about four feet tall and had round, ball-like heads with what looked like a band of mirrors around the centre and a large antenna sticking out the top. They had one short, stumpy leg that rested on a round, disc-shaped platform, something like the round, wheeled stools that you might see in a public library. Their arms were more like elephant's trunks, thick at the top and narrowing at the bottom, where they had just a few small fingers. The beings positioned themselves one in front of him and one either side. He tried to run but could not move his feet and as he flung his arms around, he found he was trapped inside some kind of bubble. At this point, the beings began to float in the air and one of them appeared to be holding a kind of hypodermic syringe which he pointed at La Ribia. He then found himself floating with them towards the huge object he had seen hovering in the field.

Once inside, La Rubia was looking down a long corridor where he noted that he could actually see the field outside, as if the walls of the craft were transparent. He felt the craft moving and the next thing he knew he was in a large, circular room, although he had no recollection of how he got there. He was lying on a table with one of the robot-like entities on one side of him, and about a dozen of them on the other. Finding his voice at last, he yelled at them as loud as he could, asking who they were and what they wanted with him. At this point they began tumbling about like nine-pins which, despite his predicament, La Rubia found most amusing.

He was shown a series of images on a large screen, including clips of himself in various situations and what he thought looked like a UFO factory, and when the show was over, he found himself outside the craft, close to the bus station. One of the beings was standing in front of him and he looked at his watch to check the time, but when he looked up, the being was gone. He then went into the bus station to check the time on the clock in there and discovered that almost an hour had passed since he had first sighted the craft. La Rubia told no one of the encounter and despite feeling unwell and suffering aches and pains all over his body, he worked his shift that day. However, he was completely exhausted and on arriving home he went straight to bed. But over the coming days he became increasingly ill, suffering from upset stomach, headaches and an itchy, burning rash that covered him from head to toe.

By the Saturday, he announced that he must quit his job as he could not work in his current condition. He finally told his friends and family about his encounter and he was referred to a psychiatrist but by now his suffering was so great, he was admitted into hospital. He was vomiting and was still covered in the burning rash and he was said to be crying like a child.

La Rubia gradually recovered physically and was later subjected to all kinds of psychological testing but was found to be completely sane and rational. Moreover, his story never changed and investigators who studied the case found him entirely believable and were certain he was telling the truth. La Rubia was not an attention seeker and would shun any approach by the media, and he eventually only told his story under duress, and because he had no choice if he was going to get the help and support he needed to recover. The episode had pretty much ruined his life and he saw no glory in it. It was a painful, traumatic experience, and all he wanted to do was to forget it and move on.

Valentich 1978

Twenty-year-old Australian Frederick Valentich was a keen flyer, as well as a UFO and science fiction enthusiast. At 6.20pm on October 21st 1978 he took off from Moorabbin airport, Melbourne in his hired Cessna 182 light aircraft to get in some nighttime flying experience. He had volunteered to collect some crayfish for the officers of his local air training corps, so he filed a flight plan that would take him over the Bass Straight to King Island, saying he would be back by 10pm.

At 7.06pm he spotted something strange in the sky, so he contacted air traffic controller, Steve Robey of the Melbourne Flight Service. He asked if he could see any other aircraft below 5,000 feet and Robey replied that he could not. Valentich then reported that he could see a large object and when asked what type of aircraft it was, Valentich replied that he could not say, but it had four bright lights around the edges. He then called in to say that the object had just passed about 1,000 feet above him and when asked to describe it in more detail, he said it was very large, but he was unable to identity it due to its excessive speed. He then asked if there were any military aircraft in the vicinity and was told that there were not.

Robey says that at that point, Valentich became anxious and distressed, and he recorded the following conversation with him.

Valentich; "It's approaching now from due east, towards me… he's flying over me, two or three times, at speeds I could not identify."

Robey; "Can you describe the aircraft?"

Valentich; "As it's flying past, it's a long shape, cannot identify more than that, it has such speed."

Robey; "Roger. And how large would the object be?"

Valentich; "It's got a green light and it's sort of metallic, all shiny on the outside."

At that point, Valentich called in to say that his engine was running rough and was missing and spluttering, then said in a raised, distressed voice, "Agh, Melbourne, that strange craft is on top of me again, hovering… and it's not an aircraft."

Robey then heard a loud cry, and a metallic sound, then silence. Valentich was never seen again and no wreckage was ever found, leaving some to conclude that Valentich's aircraft had been taken by the UFO. The official report into the incident concluded that the reason for the loss of the aircraft could not be determined.

There have been many theories about what happened to Valentich, with some saying that as a known UFO buff, he staged the incident to gain notoriety, then disappeared deliberately to lead a new life with a new identity. But this seems unlikely. Valentich was not unhappy or depressed at the time and showed no signs of mental illness. Furthermore, why would he try to gain notoriety if he were not going to be around to enjoy it? Some also claimed that he had become disorientated and confused, but Robey is adamant that he showed no signs of this during their conversation.

In support of the validity of this story, it later emerged that there had been other reports of strange craft in the sky in that area at around the same time. At 2pm on that day, a strange cloud was observed floating over King Island and a silvery-white object was seen to emerge from it. Then at 4pm, a number of independent witnesses saw two cigar-shaped objects, about the size of a jumbo jet, moving west across the state of Victoria. At around 7pm (roughly the time of the Valentich incident) several people reported seeing a green light over the Bass Straight and there were other reports of strange lights and cigar-shaped craft over Melbourne.

It is likely that we will never know what happened to Frederick Valentich, or if he really did encounter a UFO, but Steve Robey remains convinced that something terrible did happen to Valentich up there that day, and that the encounter was very real.

Chapter 7: Crashes and Landings

Many people believe that the earliest example of a UFO crashing on Earth is the famous Roswell incident in 1947, but that is not the case. There had been earlier reports of strange craft landing or crashing on Earth but without the mass-media culture that we have today, they were little-known at the time.

For example, in Alencon, France in 1790, peasant farmers were working the land when a huge metal sphere flew over their heads and crash-landed in the next field. A tall man emerged from the craft dressed in a close-fitting suit and, waving his arms, shouted something to them in a language they could not understand. It would seem he was trying to warn them, as he then ran off into the woods as the craft exploded, showering hot debris all around. It is not known if anyone was killed or injured in this event but despite extensive searches, the tall man was never found. Incidents like this went relatively unnoticed at the time, although there were others, but Roswell was about to change all of that.

Roswell

There has been so much written about the Roswell incident that it has, like the Kennedy assassination, become almost impossible to know the truth. At the time, the story attracted very little interest and was far less dramatic and sensational than it later became. In 1967, the US government appointed a UFO investigation committee to investigate known UFO cases and they asked all the leading UFO groups to list what they considered to be the best and most convincing, and not one of them mentioned Roswell.

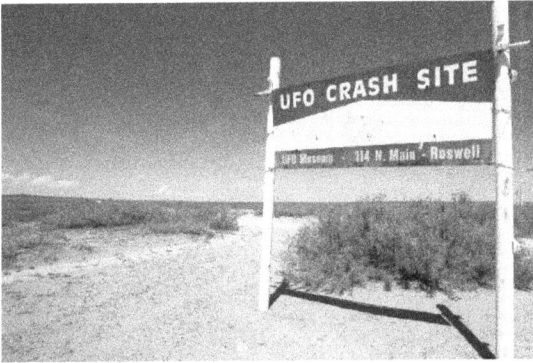

Roswell UFO crash site, New Mexico

However, in 1978, UFO investigators, Stanton Freedman and Williams L Moore tracked down Major Jesse Marcel, one of the key players in the story, and interviewed him for a TV documentary they were making called, UFOs Are Real. From that point on, the Roswell incident went from relative obscurity to being one of the biggest and most controversial UFO cases of all time. So, what really happened?

The story goes that on the 14th of June 1947, forty-eight-year-old William 'Mac' Brazel was working on a ranch 60 miles from Roswell, New Mexico, when he came across some wreckage strewn across the ground. He was accompanied by his eight-year-old son, Vernon and they both examined the wreckage which, although unfamiliar, they assumed must be from something that had fallen from the sky. They thought little of it initially but on returning to the site later, they picked up some of the pieces and drove into Roswell to inform Sheriff George Wilcox of the find. Wilcox called the nearby Army Air Force Base, home of the 509th Bomber Group, and they despatched intelligence officer, Major Jesse A Marcel to investigate. Marcel drove into Roswell and met with Brazel and Sherrif Wilcox in the Sheriff's Office to hear the story and examine some of the wreckage, which Brazel had brought with him.

Marcel was intrigued by what he saw and heard, so he called army counter-intelligence officer, Captain Sheridan Cavitt and invited him to join him on a visit to the crash site with Brazel. There, they picked up some more wreckage and took it back to the house, where they tried to piece it together and make some sense of it. However, they were still completely baffled, so they loaded the wreckage into their cars and took it back to the base. Back at the base, the wreckage was shown to base commander, Colonel Williams H 'Butch' Blanchard and he, in turn, reported it to his superiors at Eighth Army Air Force Headquarters at Fort Worth. The material was then flown to Fort Worth on a B-29 bomber.

At this point, Roswell Army Air Force public information officer, Lieutenant Walter G Haut compiled the now-famous press release and drove into Roswell to deliver it to the Roswell Daily Record in person in order to meet the deadline, and on that day, the paper carried the headline, 'RAAF Captures Flying Saucer on Ranch in Roswell Region.' The article went on to say, 'The intelligence office of the 509th Bombardment Group at Roswell announced at noon today that the field had come into possession of a flying saucer.'

Jesse Marcel with debris allegedly recovered from the Roswell crash

However, within hours of the press release, the story was already changing. Reporters were summoned to a press conference in the office of Brigadier General Roger Ramey, commander of the Eighth Army Air Force, where they were told that the wreckage was not from a UFO, but was from a weather balloon. The pieces were laid out on the floor of Ramey's office, where the press took several photos of Major Marcel holding it up for the cameras. The material displayed was really just a load of what looked like balsa wood and tin foil. The press accepted the revised version of events and after publishing this retraction in the press, the story faded into obscurity.

On revisiting the story in 1978, Stanton Freedman contacted Jesse Marcel and at first, found him evasive and reluctant to discuss the issue, but over time, he began to open up. Marcel revealed that he felt that the balloon story may have been a cover-up and said that the wreckage they found could not have come from a balloon. He described small beams of what looked like wood, but was not wood at all. It was a light, hard material that could not be bent, marked or burnt and it had some strange, hieroglyph-like markings on it. There was also a strange material like tin-foil that could not be folded and that, when crunched up in the hand, would return to its original state when released. He stated that, although similar in appearance, the material shown at the press conference was not that recovered from the site, and that they had all been sworn to secrecy by General Ramey. Marcel eventually stated that, in his opinion, the material they found that day was 'not of this Earth.'

Marcel's testimony was later confirmed by his son, Dr Jesse Marcel Jnr, a physician living in Montana. Although seemingly confused over the timing. Dr Marcel said he was shown the wreckage by his father and described the foil-like material and the small beams containing the hieroglyphs, although he failed to mention these until reminded by his father. There was another inconsistency here in that Marcel Jnr described the beams as I-shaped in profile, whereas Major Marcel said they were rectangular.

Another witness tracked down and interviewed by Freedman was Walter Haut, the author of the original press release. In 1980, Haut was reunited with Jesse Marcel for a TV documentary on the subject, where Haut revealed that he had not been aware of Marcel's recent claims that the material shown at the press conference was not that from the site. Haut remained non-committal on the subject for many years but he eventually opened up, and in 1993 he issued an affidavit stating.

"I am convinced that the material recovered was from some sort of craft from outer space."

It would appear that something did come down at Roswell in 1947 and that the authorities were unsure of its origin. There was a strong military presence in the area at the time and the public were clearly being manipulated in some way. Brigadier General Thomas DuBose was a colonel in 1947 and was General Ramey's Chief of Staff. He later said that the balloon story was a cover up and that the real wreckage was flown to USAAF Headquarters in Washington, before moving on to Wright-Patterson AFB.

But there was more to the story than just wreckage. In the late 1970's, Ufologist, Leonard H Stringfield published a paper entitled, Retrievals of the Third Kind; A Case Study of Alleged UFOs, in which he claimed that informants had told him that the US government were secretly holding the remains of extra-terrestrial beings recovered from Roswell and that at least one of these was still alive. It was further claimed by TV researcher, Robert Carr that a crashed UFO, and the frozen bodies of its crew, were being held in Hanger 18 at the Wright-Patterson AFB.

If these reports are true, they may not be linked to Roswell at all, as there were two alleged UFO crashes reported in New Mexico at that time, with a similar crash reported near the town of Aztec in 1948. This story was dismissed as a hoax, but many believe it to have been deliberately leaked out to make the Roswell story seem even less credible. Furthermore, when the Majestic-12 papers were declassified in the mid 1980's, one document revealed that President Truman had ordered the formation of a group to oversee studies of UFOs and alien bodies recovered from Roswell, although this document was later declared to be a forgery.

In a further twist to the Roswell story, in 1997 retired US Army Lieutenant Colonel Philp Corso released his controversial book, The Day After Roswell. Corso had served many years in Army Intelligence and had once been on the staff of President Eisenhower's National Security Council. In his book, Corso claims that in 1961 he was appointed head of the Pentagon's Foreign Technology Desk, under the command of Lieutenant General Arthur Trudeau.

Corso says that in this role, he led a double life in that he was tasked with progressing conventional weapons development programs as well as covert projects relating to Roswell. He claims that the term, Foreign Technology included alien as well as terrestrial technology, and that as head of his new role he inherited a top-secret filing cabinet containing documents and artifacts from the Roswell crash.

These were then used to reverse-engineer alien technology, and this was done by filtering this alien technology through selected government contractors like Rockwell, IBM and the Hughes Corporation, to be reverse-engineered and used in the development of advanced weapons systems.

He claims that this collaboration led to the rapid development of new, advanced technology and applications such as the integrated circuit, fibre-optics, lasers and night vision technology, to name but a few. He says that this was very much a feature of the cold war as the Soviet Union were aware of an alien presence on Earth, and that the Soviets vied with the Unites States to acquire and develop this technology. Moreover, he says that alien technology was used extensively in President Reagan's Strategic Defence Initiative and that the SDI project was developed as much to counter a potential alien threat as to defend against the Soviet Union.

In his book, Corso goes on to claim that alien bodies were recovered from the Roswell crash and that the US government were harbouring at least one live extra-terrestrial of the 'grey' type. He says that whilst at a base in New Mexico he was shown a crate that contained a glass box, inside which was an alien body, apparently dead and suspended in some sort of liquid. The description he gives is of the typical grey alien type, about four feet tall, with thin arms and legs and a huge, bulbous head with large almond-shaped eyes. He said he was so shocked and horrified by this, he was almost sick.

The Roswell crash story has had, and still has, many critics. Some claim that whatever came down in Roswell was very much man-made and that the reason the government covered it up was to hide the truth about some secret military project being developed in the area. One potential candidate put forward for this was a top-secret project called Project Mogul. Mogul was a highly classified project designed to detect Soviet nuclear tests by sending sophisticated monitoring equipment into the stratosphere using high-altitude balloons. These balloons and their cargo were known to only a few senior personnel, so the components that went into making them would have been unfamiliar to those not in the know. This explanation, however has long been rejected by UFO enthusiasts and by many mainstream scientists as a very weak and unconvincing cover-story. Over the years, and with the passing of many of the key witnesses, it is now almost impossible to know what really happened at Roswell unless the authorities chose to tell us. But whatever it was, it was not an isolated case. The Roswell incident is one of the most famous UFO stories of all time, but it is by no means the only account of a UFO crash.

The British Roswell

Following a spate of UFO sightings in the north of England in 1974, Britain was to get its own Roswell. At about 8.30 on the evening of 23rd January that year, the residents of the villages around Llandrillo, North Wales were suddenly startled by two loud explosions that could be felt through ground and shook the houses. Some reported seeing a bright flash in the sky, followed by an eerie afterglow. Believing there may have been a plane crash, the emergency services were alerted and nurse, Pat Evans and her daughters drove up to the mountain to see what had happened and to offer assistance. When she arrived at the scene, she saw a huge spherical object that seemed to be resting on legs and was giving off an orange glow, that pulsated and changed colour. It had taken some time to reach the remote site but when she arrived, Nurse Evans found the authorities already there and that the area had been declared off-limits. Helicopters were seen flying over the area as if looking for something and sometime later, mysterious visitors arrived to question witnesses. Some locals later said that within a few hours of the crash, an army truck had been seen coming down from the site with a huge crate on the back.

The incident languished in relative obscurity until resurrected by UFO investigators, Jenny Randles and Nicolas Redfern. They claimed to have found government documents relating to Operation Photoflash, which involved the retrieval of a downed UFO from the sea off the North Wales coast. The story was that a UFO had zapped a Royal Navy ship, killing some of the crew, and was shot down by military jets before later being recovered. The investigators went on to suggest that the UFO that came down in the Welsh mountains did not crash, but landed, and that the occupants conversed with military personnel. There were also claims that alien bodies had been recovered and taken to a secret government facility in Porton Down.

Another researcher, Scott Felton investigated the case further and his findings supported those of Randles and

Redfern. Felton was interviewed by researchers for the TV documentary series, Britain's Closest Encounters, but when the program was aired, his testimony had been edited out and replaced by the official conclusion that the crash had been nothing more than an earth tremor.

The Brazilian Roswell

The story behind what has become known as the Brazilian Roswell is a strange and complex one. Against a background of UFO sightings around Varginha, Brazil, on the night of 19th January 1996 an American satellite spotted an unidentified object flying in the direction of Varginha. At around the same time, amateur video footage showed strange, white objects hovering over and around the town at various heights, sometimes close to the ground and at other times, high in the air.

The following day, the Varginha fire brigade received reports that strange creatures had been seen on the outskirts of the town and when they arrived on the scene, they found that the military were already there. The fire-fighters then collaborated with the army to search for the creatures, one of which they found hiding in long grass, apparently injured. The creature appeared timid and docile, so was easy to capture. Witnesses said they heard three shots, then soldiers emerged carrying a sack with something alive and moving inside.

Later that same day, three local girls spotted a strange creature crouching behind a wall. The girls screamed and, shocked and startled, the creature turned to look at them.

They described the creature as being short, with soft brown skin and long, skinny arms. It had big, red eyes and it seemed to be in pain. At 10-o-clock that night, the military came across another creature lying in the road. This one, too, was alive and was taken to the local hospital, but died soon afterwards. The hospital was sealed off and a post-mortem was carried out on the body. The creature was said to have a long tongue in its slit-like mouth and the eyes had no pupils. The arms were long and thin, with protruding veins and the body gave off a strong smell of ammonia.

Once the autopsy had been completed, the body was flown to Sao Paulo University, where it was placed in the care of Dr Badan Palhares, head of forensic medicine. When asked about this by one of his students, Palhares denied any involvement but added cryptically, "Ask me again in fifteen years." Even the most ardent UFO enthusiasts have had to admit that there is very little evidence to support this story. Some elements may be true, but there has been no proof of aliens running around the town and the official investigation declared that the so-called alien seen by the girls was actually a homeless vagrant.

The Rendlesham Forest Incident

The Rendlesham Forest incident remains one of the best-known and best documented UFO incidents of all time, and is still hotly debated today. Rendlesham Forest, in Suffolk, lies between the former USAF bases of Bentwaters and Wood Bridge and it was here that, over the Christmas holidays in 1980, this momentous event occurred.

In the early hours of 27th December 1980, two USAF security guards, Sergeant Jim Penniston and Airman John Burroughs, were patrolling the perimeter fence of their base in Rendlesham forest when they saw a bright light shining through the trees beyond the perimeter. Thinking it might be a downed aircraft they requested, and were given, permission to go into the woods to investigate. They took their flashlight and made their way towards the light and as they got closer, they could not believe what they were seeing.

Resting on the forest floor, or hovering just above it, was a strange, triangular or pyramidal-shaped craft, just twenty metres away. The object was two or three metres across and about two metres high and gave off a blueish, white light that seemed to pulsate and change colour. The underneath was lit by a blue glow and on top was a flashing red light. Both men were experienced airmen with a good knowledge of aircraft types and recognition, but this was like nothing they had ever seen before.

As Burroughs stayed back to cover him, Penniston gingerly approached the craft. The air was electrically charged and he said that his movements were laboured, as if he were walking through water. When he got to within a few feet, his movements became easier and the craft changed colour and appeared black and smooth, like black glass, and appeared to have no visible seams or other features. Moving closer still, Penniston noticed some strange, hieroglyph-like symbols on the side of the craft, and he reached out to touch them.

As his fingers made contact with the symbols, he felt a sudden jolt through his body and he felt he was being subject to some kind of sub-conscious download. He saw an array of ones and zeros flashing through his mind and he was dazzled by an intense, white light. He pulled back his hand and moved away, but the numbers and symbols kept racing through his head. The craft then rose from the ground and moved off, leaving Penniston and Burroughs completely shocked and mystified. Meanwhile, Sergeant Penniston could not get these symbols out of his mind, so he recorded them in his notebook, where they remained, forgotten for almost thirty years.

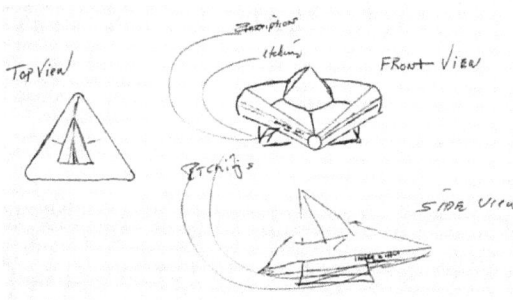

A sketch of the Rendlesham UFO done by witness, Staff Sergeant Jim Penniston

The two airmen reported the incident and, although still mystified, considered the matter closed. But two nights later, the lights returned. Personnel at the base reported seeing lights in and around the forest once again, and these were seen to hover and manoeuvre over the trees. On seeing this, deputy base commander, Lt Col Charles Halt led a team of men out into the forest to investigate. On entering the woods, Halt and his men saw a strange red light through the trees which was flashing and was described by Halt as looking like a blinking eye. It was Halt's habit to carry a tape recorder on such occasions and the recording he made that night is truly shocking. On the tape, the men can be heard discussing the strange events and are clearly confused and afraid. On approaching the site of the landing seen by Penniston and Burroughs on the 27th, Holt can be heard saying.

"We're getting a high radioactive reading... Indications of a heat source coming out of that centre spot.'

There were three indentations in the ground from the craft's landing legs and these gave off a high radioactive signal. They also saw some damage to the trees around the landing site, with branches broken off the trees, and some of the vegetation had been burnt.

At that point, their attention was again drawn to the red, pulsating light that now seemed to be moving towards them. Halt's tape recorder captured the following conversation.

"There it is."

"Yeah, I see it, too."

"What is it?"

"We don't know, sir."

"There's some sort of strange, flashing light, and it's coming this way... It's definitely coming this way... Pieces are shooting off it... There's no doubt about it, this is weird."

The light began to move away and as it did so, a second UFO appeared above it and fired an intense beam of white light onto the ground in front of the men before it, too, disappeared.

Soil and vegetation samples were taken and, along with the video and audio evidence, were submitted with Lt Col Halt's report, and it is believed that these were later sent back to the United States. Halt also submitted a detailed report to the UK Ministry of Defence, but heard nothing back from them. A few days later, the men were interviewed by mysterious men in suits who told them they had been mistaken and that the event had never happened. They were to discuss this with no one, on pain of court martial.

Many years later, now retired Jim Penniston felt free to talk about the incident and whilst giving one of his rare interviews on the subject, he showed the interviewer the drawings of the symbols in his notebook. On examining these, the researcher noticed the pages containing the ones and zeros and asked Penniston what they were, and he said he did not know. He had just written down what came into his head shortly after the incident, then thrown the notebook into a drawer and forgotten about it. The researcher pointed out that the numbers looked very much like binary code, of which Penniston had no knowledge, and suggested that they could possibly be decoded. Penniston agreed to that proposal, and the note book was given to a binary expert to decipher. The analyst was given no information about the origin of the coding and was told nothing of Penniston's background or experiences.

The results were astounding. The binary expert was able to decipher a message from the coding, which made no sense to him, but which read.

'Exploration of humanity. Continuous for planetary advance. Eyes of our eyes. Origin 8,100.'

The message went on to give the co-ordinates of several sacred sites around the world, including the pyramids at Giza and the Nazca Lines in Peru. Furthermore, one of the locations given pointed to an empty piece of ocean in the North Atlantic and at first, that seemed strange. But further study has revealed that there was once a land mass at this precise location, which is shown on ancient maps and is known as Hy Brazil, and this is often referred to as the other Atlantis.

Despite all of this, the UK government continued to show little interest in the Rendlesham Forest case, stating that it was of no defence significance. This prompted former Chief of Defence Staff, Lord Hill Norton to declare that, "Either the Americans, and indeed, the deputy base commander, were hallucinating, or they really believed that something had landed there and they had taken photos and records of it."

Some sceptics have tried to argue that the lights the men saw that night were from the nearby Orford Ness Lighthouse, but that theory comes nowhere near to explaining what happened. Although operating at the time, Orford Ness was well-known to the men at the base and they have all stated quite categorically that what they saw could not possibly have been the lighthouse.

UFO Lands in a field 1981

French farmer, Renato Nicolia, was building a shelter at the back of his house on the afternoon of January 8th 1981 when he heard a strange whistling sound. He looked around for the source of the noise and, realising it was coming from above, he looked up and was startled to see a strange, spherical object descending into a nearby field. The object was small, only a few feet in diameter, and looked like two bowls turned one on top of the other, with a sort of crown on the top. It was metallic grey in colour and had small landing legs on the underside. Renato noticed that the object appeared to be falling, rather than making a controlled descent, and it hit the ground with a thud.

After a short time, Renato approached the object to get a better look and as he did so, it made a high-pitched whistling sound, then shot up into the air at high speed. As it rose up into the sky, he noticed an opening on its underside, although there was no smoke or flame coming from it. He also noticed that the craft had picked up a small amount of earth in a kind of claw-like device. On examining the area where the object had landed, he saw indentations in the ground where the craft had stood.

He called the police, and they took photos and earth samples, which were sent to Toulouse for analysis. The case was then handed over to the GEPAN, the French government's official UFO investigation body. Results of the analysis of the material showed that the earth had been crushed by a very heavy object and that the ground had been heated to between 300 and 600°c. Furthermore, the soil showed excessively high levels of iron and phosphates, along with very high levels of zinc, and plants in the vicinity had been bleached of 50% of their chlorophyl. Various theories were put forward for these environmental changes, but none could be satisfactorily explained.

Landing in the Lake District, 1988

On the night of September 22nd 1988 plumber, Noel Harrison and his girlfriend, Maxine Williams were driving in their transit van along the M6 motorway on their way to visit Noel's cousin in Grayrigg. They had made the journey many times before and were familiar with the roads around Kendal and Windermere, but as they turned off the main road into a deserted lane, they noticed that the road ahead was blocked by what appeared to be large boulders. As the countryside around was flat, with no high ground or mountains, they were puzzled as to where the boulders could have come from, so they stopped the van and got out to have a look around. On closer inspection the boulders appeared to be two-dimensional and as they approached them, they seemed to disappear. They would also disappear if looked at from an angle, and could only be seen from directly in front. Mystified, Maxine found the whole experience quite spooky and she urged Noel to take another route, but curiosity had got the better of him. He drove the van slowly towards the obstructions and as he drew closer, they disappeared, and the van passed harmlessly through them.

Then, just as they thought their strange ordeal was over, they rounded a bend in the narrow road and saw, in a clearing to the left of the road, a huge, saucer-shaped craft standing on four legs. Beneath the craft were two figures, approximately 7 feet tall and wearing large helmets, who seemed to be working on it in some way. Noel continued to drive slowly past and the figures turned to look at them, as if startled. A beam of light shot out from the craft and hit the van, and Noel accelerated away as fast as he could.

When they arrived at their destination in Grayrigg, Noel told his cousin what had happened and was informed that a strange disc had been seen flying erratically over the nearby village of Middleshaw that same night. They returned to the site the next day and saw four indentations in the ground where they had seen the craft the night before, although they could see no sign of the boulders. The police were informed and carried out an investigation but were unable to offer any explanation for the event, although they were convinced that the witnesses were telling the truth. The incident was also investigated by UFO researchers, who suggested that the craft may had experienced difficulties and had landed in a secluded spot to carry out repairs. The boulders, it was theorised, were optical illusions, put there to deter any unwanted attention.

Chapter 8: UFO-Related Phenomena

There are a number of strange phenomena often associated with UFOs, from the blood-sucking Chupacabra of South America to the Mothman of Virginia and the infamous Bigfoot. UFO sightings are frequently accompanied by other strange occurrences and events that defy explanation and, if real, may very well be linked to UFO activity. These UFO-related phenomena fall into many different categories and are very varied in nature, and we could not cover them all in this book. But below are some examples of such cases in the three most prominent and best-known categories of UFO-related phenomena.

Cattle Mutilations

One of the earliest recorded cases of cattle mutilation occurred on September 8th 1967. On that night a three-year-old pony named Lady failed to return to her corral, and on searching his land, ranch owner, Harry King found her dead in a field. The flesh had been stripped cleanly from her head and neck with remarkable precision and there was no evidence of any intruders in the area. There was no blood at the scene and there were no signs of predator activity. Furthermore, on the same night, some local residents had reported strange lights in the sky, and some began to suspect a link between the two incidents.

Thus began a spate of similar cases that persisted well into the 1990s and beyond, although reported cases are far less frequent today. In his book, Messengers of Deception, UFO researcher, Jaques Vallee noted that 23 states were affected by the phenomena and that in the 18 months leading up to January 1977, there had been more than 700 cases of cattle mutilation in 13 western states alone. Reports of cattle mutilation have often been accompanied by sightings of strange lights in the sky and apart from a few cases in the UK and in Australia, the phenomenon has been mainly confined to the United States.

There are many mysteries surrounding the phenomenon of cattle mutilations and up to now, no one has been able to establish who is doing them and why. In a typical case, cattle, and sometimes horses, will be found dead in a field with large parts of their bodies removed with surgical precision, suggested by some to have been cut using lasers. The blood from the animal will often be drained but there is no evidence of blood on the ground and there are no tracks or evidence of movement around the victim, as if it had been lifted and dropped from above.

On one occasion in September 1994, Larry Gardea was tending his cattle on his ranch in Chaco Canyon, New Mexico when he heard a strange humming sound. The herd began to run away from the noise and to his utter shock, Gardea saw three calves being dragged, bellowing, towards a beam of light. He fired his gun and the humming noise ceased, so he ran over to inspect the calves, who were now lying on the ground. When he got there, he found that one was dead and horribly mutilated, one was dead but had no visible wounds and the third was nowhere to be seen. Gardea reported this to the local sheriff, who came out to investigate. The two arrived on the scene, and on examining the mutilated calf they found that its jaw had been completely stripped of flesh on one side, revealing the pure white jawbone underneath. Its tongue had been cut out and its reproductive organs removed. The wounds were very clean and precise and there was no blood around the animal. It was thought that the second calf died of fright, and the third one was never found.

Cattle mutilations have been investigated by many people over the years, but perhaps most notably by researcher and documentary film maker, Linda Moulton Howe, who is now a leading authority on the subject. On discussing her book, Alien Harvest in an interview with the BBC, Ms Moulton Howe described a typical cattle mutilation as. "Ears missing, eyes missing or the flesh around the eye removed, and the jaw flesh is usually removed, too. You are usually left with about half the animal. The tongue is removed by a cut so deep that it goes all the way down to the larynx, and in many cases not only is the larynx taken, but a large part of the trachea, too."

In her book, Ms Moulton Howe surmises that, "This can't be predator activity. All the tissue is dry and there are no tracks around the animal." She then goes on to ask, "How could a 2,000lb bull in Colorado end up on its back with its horns dug six inches into the ground and its forelegs up in the air?"

Incredible as it may seem, the popular belief amongst some Ufologists is that cattle mutilations, along with some of the intrusive surgery experienced by many abductees, are part of an alien program of genetic engineering. The theory is that aliens, probably the greys, supervised by the Nordics, are removing organs and other bodily parts and cells from both humans and cattle, to use in the creation of some kind of hybrid race, perhaps because their own race is facing the risk of extinction through some unknown genetic defect or deficiency.

Although possible, this hypothesis is hard for most people to accept. But there can be no denying that something strange has been happening to cattle, especially in the United States, for decades. Pictures of cattle mutilations are hard to look at and show, in a very graphic and disturbing way, just how distressing this phenomenon is. The wounds on these poor animals are very neat and precise and have clearly been executed with great skill and care. It is possible that they are hoaxes, or even that they may be connected to some kind of ritualistic devil-worship, but both of these theories seem unlikely.

Crop Circles

Like cattle mutilations, crop circles are another mysterious phenomenon often associated with UFOs. There has been much debate about how these patterns in crops are created and we often see reports of strange lights in the sky before crop circles appear. This has led some to believe that crop circles are some kind of communication mechanism, used by aliens to relay messages and ideas to us here on Earth, using the universal languages of geometry and mathematics.

Reports of patterns appearing in crops are not new and have been reported since the late middle-ages. The Mowing Devil, a famous wood-cut illustration dated from 1678, shows a devil-like creature making patterns in crops using what looks like a scythe, and is the earliest known account of the appearance of what we would now call a crop circle. There have been other, occasional reports of crop circles in the 1940s and 50s, but they became far more prevalent from about the 1970s onwards. These early reports were of basic, swirling patterns and were mainly confined to the Wiltshire area of England, close to Stonehenge and Salisbury Plain. However, over time the patterns became infinitely more complex and much larger in scale. One such pattern appeared overnight at Milk Hill in Wiltshire and consisted of 420 accurate circles forming a complex spiral-pattern design that, although difficult to see from the ground, appeared perfectly symmetrical when viewed from the air.

Crop circle formations have been reported in many countries around the world, but they appear to be far more prevalent in the English countryside. Independent researcher, Lucy Pringle has been studying crop circles for many years and has theorised that they are created by extra-terrestrials, as they display some strange and inexplicable features and characteristics. She has found that on entering a crop circle, some people report strange feelings of unease and even nausea. Ms Pringle also found that most crop circle events occur in chalk-bed areas or in areas with vast amounts of underground water and that whatever force is used to create the circles, it is thrown vertically from above, creating a huge electrical discharge. This results in micro-wave activity which softens the base of the corn stalks, causing them to fall over but not break.

Research has shown that in a genuine crop circle, the outside of the plant stalks are lengthened and the joints in the stems (the nodes) are stretched and bent at an angle, but not broken. Sometimes these bends in the stalks appear six inches from the ground, and so could not have been created by treading on them. Furthermore, the stalks form small holes, as if heated from the inside, and the molecular structure is changed, with seeds showing signs of micro-wave damage from the inside out. None of this would be possible if the circles were created artificially by pranksters.

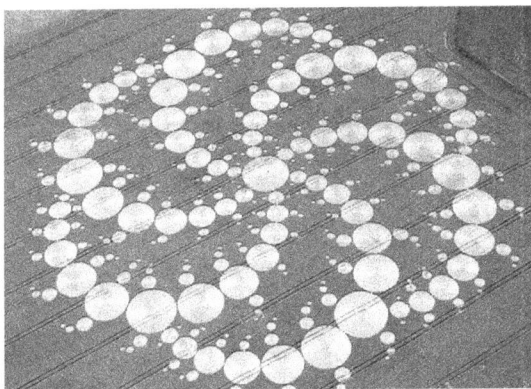

Crop circle in a Wiltshire field

Sceptics have long claimed that crop circles are just a huge hoax and are man-made, but that seems unlikely in at least some of the cases. Whilst a large number of crop circles are clearly hoaxes, some are not. In 1991 two practical jokers,

Doug Bower and Dave Chorley famously declared that they had been making patterns in the fields since at least 1970, and that they had created many of the famous formations that had baffled researchers for so long. They claimed they could create such a pattern very quickly and demonstrated how they did this by trampling down the crops using a plank held by a piece of string. But even if this does account for many of the famous crop circle events, it cannot explain them all. It would be impossible for two people to create some of the larger, more complex formations in one night, without being seen, and they could in no way replicate the physical and biological changes in the crops that we see in genuine crop circle cases.

Reports of crop circles are now quite rare, and the phenomenon has all but died out, but there are still some that remain unexplained. That is not to say that they are created by extra-terrestrial, just that we don't know how they were created or why.

Men in Black

Of all the many attributes often associated with UFO encounters, the Men in Black phenomenon is perhaps the strangest and most intriguing. Many witnesses have claimed that shortly after experiencing a UFO encounter, they have been visited by mysterious men dressed all in black and often driving distinctive black cars (although they sometimes arrive on foot, or even in black helicopters). Men in Black reports were more common in the 1960s and 70s and are less frequent today, but they do still happen and although many can be easily dismissed as hoaxes, many more have defied any explanation.

There are variations to these stories, but the general pattern is that shortly after experiencing a UFO encounter, the doorbell rings and the witness is confronted by a complete stranger they have never seen before (sometimes there are more than one, and they may occasionally be accompanied by a woman). They are generally dressed in immaculate black suits, a homburg-style hat and dark glasses and will often wear grey gloves. Their behaviour and speech seem strange and their terminology and language appears to be old-fashioned, and even corny, often described as being like a script from a bad B movie. Their movements are often described as robotic and, most bizarrely of all, many seem to be wearing make-up and lipstick. Having gained the attention of the witness, they proceed to warn them not to discuss their encounter with anyone, often making threats that in most cases come to nothing.

A 'Men in Black' visitation following a UFO sighting

Albert Bender

One of the most famous cases of the Men in Black phenomenon is that of Albert Bender, Director of the International Flying Saucer Bureau. In 1952, Bender had been experiencing some strange encounters with mysterious men dressed in black who visited him at his home, in the street and even in the cinema. They were telling him that they were unhappy with the activities of his very popular Flying Saucer Bureau and urged them to curtail their research, or terrible things would happen. Bender ignored these threats and announced that, what he called a startling announcement, would be forthcoming in the July edition of the Bureau's magazine, *Space Review*.

However, as Bender was resting at home two weeks later, three figures materialised in front of him and he suddenly felt an overwhelming sense of dread and experienced nausea and a sudden headache. Bender later said.

"As the figures became clearer, I could see that all of them were dressed in black clothes, just like the men I had seen in previous encounters. They looked like clergymen but wore hats similar to the homburg style. Their faces were not discernible, as their hats partly shaded them, then the feeling of fear soon left me. The eyes of all three figures suddenly lit up like flashing bulbs, and they were all focussed on me. They seemed to burn into my very soul as the pain above my eyes became almost unbearable. It was then I sensed that they were conveying a message to me by telepathy."

The visitors allegedly told Bender that he was on the right track with his investigations on UFOs, but that he should not publish his forthcoming article. They told him that flying saucers were, indeed, from outer space (specifically, the planet, Kazik) and that the aliens were already with us, here on Earth. He was informed that great harm would come to him and others if he published the article, and that people have died in defiance of similar instructions. This time, Bender took heed. He withdrew the article, ceased publishing the magazine and even disbanded the Flying Saucer Bureau. He then released the following statement.

"The mystery of flying saucers is no longer a mystery. The source is already known, but any information about this is being withheld by orders from a higher source. We would like to print the full story in Space Review but because of the nature of the information, we are sorry that we have been advised in the negative. Furthermore, we advise that all those who engage in flying saucer work to be very cautious."

Robert Richardson

Another strange case is that of Robert Richardson from Toledo, Ohio. In 1967, Mr Richardson was driving his car when he hit something in the road. He was not sure what it was, as it immediately vanished. He stopped the car and got out to investigate and found a strange lump of metal in the road, although he was sure this was not what he had hit. He found the whole thing quite puzzling, so he picked up the lump of metal and sent it, along with his story, to the APRO (Aerial Phenomena Research Organisation) for analysis.

Shortly after, Mr Richardson was visited late at night by two men dressed in black suits, who quizzed him about the incident for twenty minutes or more. They showed him no ID (as he didn't ask for it) and he said he felt no hostility from them at all. He felt calm and composed and was more than happy to discuss the incident with them. They thanked him for his time and left in the car they had arrived in. This was a 1953 black Cadillac, 14 years old but in mint condition. This intrigued Mr Richards and prompted him to check the licence plate, which was found to be false.

One week later, two more dark-complexioned men in black suits arrived, this time in a contemporary Dodge. Both seemed foreign, but one spoke with a particularly good English accent. The visitors this time seemed more threatening and tried to convince him that he had seen nothing odd on the night he had hit the object. Mr Richardson tried to argue back but they were most insistent, and after applying considerable pressure, they asked him for the metal object he had found. When he said he no longer had it, they said, "If you want your wife to stay as pretty as she is, you had better get it back."

This may seem odd and might even make the story less believable, but many MIB (Men in Black) witnesses say that the visitors used strange, almost comical terminology like this. However, as is often the case, these threats came to nothing and Mr Richardson heard no more from his mysterious visitors.

In the North Yorkshire seaside town of Scarborough in 1968, sixteen-year-old schoolgirl, Adele, was at home alone when she heard a knock at the door. On answering it, she was confronted by a tall man wearing a black suit, white shirt and black tie, and what she described as a pork pie hat. He had a very ruddy, clown-like complexion and wore a creepy, beaming smile. However, she later said that despite this unnerving detail, she felt more startled than afraid. After grinning at her for what seemed like ages, but was really only a few seconds, the man's whole body jerked and he said, "Do you have insurance, is it now?"

Adele did not know what to make of this question and could make no sense of it at all. After taking a moment to compose herself, Adele politely told the visitor that he should speak to her father about that, but he was not at home right now, so could he call back later. At that point, the man began to sweat profusely and removed his hat to wipe his forehead with the back of his hand, revealing a completely bald, pure white head. It was then that Adele realised that the man's ruddy complexion was actually very badly applied stage make up.

The stranger then asked, "Can I see a glass of water?" and fearing that he may have been about to faint, Adele went into the kitchen to fetch one. As she did so, the man followed her into the kitchen and although she still felt no fear of him, she did notice some peculiar features. He walked in small, jerky steps with his head thrown backwards, which she found quite amusing. She noticed, also, that his trousers and coat sleeves were too short and revealed white, hairless arms. But perhaps most peculiar thing of all was that his shiny black shoes were on the wrong feet.

Adele showed the man back into the lounge and when she returned with the water, she found him standing in front of the fireplace, staring at a carriage clock on the mantlepiece, which she told him was her father's. She handed him the glass of water and to her surprise, he just held it up to the light and looked at it, making no attempt to drink it. At that point, she realised that he had actually asked if he could see a glass of water, and that was exactly what he was doing. The man gave her back the glass of water and turned back to look at the clock, tapping it repeatedly with his finger, and saying, "Your father, your father. His time, his time."

The man then turned around and said, "Watch the lights," before heading, with some difficulty, for the front door. Realising that the man was leaving, Adele rushed to the door to open it, and the man walked awkwardly out without uttering another word, or even looking at her. As soon as she had closed the door behind him, Adele rushed to the window to watch the man walk down the street, but she could see no sign of him. She had a clear view down both ends of the street and there was nowhere he could have turned off, but he was gone.

Dr Hopkins' MIB Encounter

In 1976, Dr Herbert Hopkins of Maine was working on a UFO case in his capacity of a professional hypnotist, when he received a strange telephone call. A man claiming to be the Vice President of the New Jersey UFO Research Organisation called him and asked if he could call round to discuss the UFO case with him. Dr Herbert agreed, and almost immediately after putting the phone down, the man turned up on his door. He had arrived on foot, which was strange, as there were no mobile phones in 1976 and the nearest call box was some distance away. Dr Hopkins was alone in the house at the time, as his family were all out, so he invited the man in for a coffee.

On entering the house, Dr Hopkins was able to get a good look at the visitor and his first impression was that he looked like an undertaker. He was smartly dressed in a black suit, white shirt and a black tie and he wore shiny black shoes. He also wore a pair of grey gloves and a hat, but the most striking thing about him was his face. His skin was pure white and he had no eyebrows or eyelashes. Moreover, his lips were bright red, and when he wiped his mouth with the back of his hand, a smear of red lipstick came off onto his glove.

Having listened to Dr Hopkins' account of the UFO case he was working on, the man instructed him to erase all the tapes of the hypnosis sessions with the witness. He told the doctor that he knew he had two coins in his pocket, which was correct, and he asked him to take one of them out and place it on the table. He told him to watch the coin and not take his eyes off it and as he did so, the coin disappeared right in front of him. The man then said, "Neither you, nor anyone else on this planet, will ever see that coin again."

As their discussion continued, Dr Hopkins noticed that the man began to speak more slowly, almost as if he was losing power in some way. After a few minutes of this, the man rose, very slowly, saying, "My energy is running low, I must go now. Goodbye."

The man then moved jerkily towards the front door and on passing through it, walked towards a bright blue light on the driveway, which Dr Hopkins thought might have been a car, although he could not actually see the source of the light.

All of this felt quite normal to Dr Hopkins at the time, and it was only on later reflection that he realised just how strange the whole experience had been. He began to feel increasingly uneasy about it, and he ceased his UFO investigation and erased all the tapes. Moreover, on further investigation, he found that the New Jersey UFO Research Organisation did not even exist. He tried to put the episode out of his mind and move on, but the nightmare did not end there.

Shortly after this disturbing encounter, Dr Hopkins and his family received a telephone call from a man who asked if he and a colleague could come and meet them at their home. They agreed, but once again, it was only after the event that they realised what a strange request this was. The strangers, a man and a woman, duly arrived and seemed normal enough, but closer inspection revealed some striking oddities. They were both aged about thirty but wore very old-fashioned clothes that looked like new. Their movements seemed laboured and somewhat robotic and they leaned forward as they walked, taking small, careful steps. They were offered a drink, which they accepted, although they did not touch them, which their hosts thought odd, but it was their behaviour that was strangest of all.

As they questioned the Hopkins family, the visitors spoke in a strange, disjointed way and they asked some very personal and embarrassing questions. The man began fondling the woman in a clumsy, awkward way and asked if this was what people did and if he was doing it right. The personal questions continued and when Mr Hopkins briefly left the room, the man asked Mrs Hopkins if she had any nude photos of herself, which she found shocking and quite disturbing.

Deciding they had had enough, Mr and Mrs Hopkins were about to ask their guests to leave, when they made a move to go anyway. But there was a problem with the man. He stood up but seemed to have difficulty moving and although the female tried to help, she could not manage him on her own, so she asked Dr Hopkins for assistance. The couple then left, without uttering another word. On discussing these events later, Dr Hopkins said that on both occasions the visitors seemed to experience some sort of power loss, as if they were robots, which might also explain their awkward speech and jerky movements.

The MIB and the Figure Skater

Award-winning figure skater Collette Peters was convinced that she had seen at least two UFOs as a teenager and has always related these events to a terrifying experience she had some time later. Whilst walking down a street in New York one afternoon, she encountered a strange looking man emerging from the crowd towards her. She has described this experience as follows.

"I was walking down one of the main avenues when I started to get a bizarre feeling. It was just pure evil. I suddenly felt very frightened and concerned for my safety. When I looked ahead of me, I saw a tall man walking towards me, but he seemed out of place and quite odd. His skin did not look normal and I could see no facial hair. In fact, his face was quite plastic-looking, smooth and shiny with no marks or blemishes at all. He didn't come directly up to me and threaten me or anything, and I don't think he conveyed any message to me. But he had a sinister presence and he emanated a disturbing feeling of evil, that's the only way I can describe it. I felt harassed, if you can put it that way, and I didn't know what to expect. I don't feel it could have been a figment of my imagination, because of the physicality of it and the feelings that I felt. I am not talking about emotional feelings; I am talking about the hair actually rising on your skin. It was something I will never forget."

The MIB and the Healthcare Worker

One night in October 2014, a healthcare worker in Yorkshire heard noises outside her house and, thinking that a fox was attacking her chickens, she went out with a torch to investigate. As she approached the pen, she noticed a figure standing in front of her and, shining her torch in that direction, she saw that there were, in fact, two of them. They were well over six feet tall and were wearing black business suits and hats. Their faces were smooth and almost featureless, with no nose and just small slits for a mouth. They asked her to please turn off the torch as it was hurting their eyes and as the outside light was now on and illuminating the area, she complied.

Using old-fashioned language and with a quiet, but firm voice, they began to question her. They asked if she had seen anything strange in the area and she replied that she had not, which seemed to satisfy them. They thanked her for her time and as they turned to walk away, she noticed with some amusement that they had very long feet.

Considering it a strange, but isolated incident she thought little more about it until sometime later, she was visited again. Whilst having a cigarette outside at work, she was looking down at the floor when a pair of black, shiny shoes came into view. She looked up and saw a man, like the ones she had seen by the chicken pen who, again, asked if she had seen anything strange. She repeated that she had not and he again thanked her and turned to walk away. His face was featureless and plastic-looking and his movements were somewhat jerky. Acting on impulse, she looked down at his feet but on this occasion, they looked quite normal.

Reports of mysterious Men in Black have come from all walks of life and from many parts of the world and, although often dismissed as hoaxes or fabrications, many of the victims are left distressed and traumatised by the experience. One witness told how he was in a compartment on a train with no corridor, when he noticed a man dressed in black and wearing dark glasses, sitting opposite and staring at him. He did not recall where the man had boarded the train, but he was relieved when he finally arrived at his destination and was able to alight and leave the man behind. He thought he had seen the last of this menacing stranger, but as he pushed his way towards the station exit, the man suddenly appeared from the crowd. He leaned close to the witness and said, "Can you spare your life?"

The bewildered passenger had no idea who the man was or what he meant by this, but although he did not see him again, the incident troubled him for many years to come.

Whether real or not, the mysterious Men in Black have become a key element of the UFO phenomenon. They are almost always connected with UFO sightings, but their purpose and intentions remain unknown.

Chapter 9: The Moon and Mars

Mars

Mars is the most Earth-like planet in our solar system, but at a distance of 142 million miles from the Sun, it is still very different. Mars is bitterly cold, with temperatures ranging from -23° to -137° and it rotates on its axis once every 24hrs and 39mins, making a Martian day slightly longer than a day on Earth. Its very low atmospheric pressure (made up mainly of carbon dioxide) means that it cannot sustain liquid water on the surface, but evidence suggests there may be large quantities of water underground. The planet does have north and south polar icecaps but, like the atmosphere, these are made up of mainly carbon dioxide. In short, it is a very inhospitable place, which many believe could not possibly sustain life. But this is not necessarily the case.

Deep canyons and ravines on the surface of Mars show tell-tale signs of water erosion, which suggests there must have been liquid water on Mars sometime in the distant past. The fact is that most scientists now agree that Mars once enjoyed a warmer period at some point in its turbulent history and may once have been ecologically very similar to Earth. This warmer period is thought to have lasted around a million years, not long enough for intelligent life to evolve, but easily long enough to develop micro-biological life forms of some kind. And we may have already found it!

In August 1996, a group of researchers working at the Johnson Space Centre announced to the world that they had found positive proof that life once existed on Mars. On slicing through a meteorite found in the Antarctic and known to have come from Mars (ALH84001), they found small, tube-like structures that looked like micro-fossils. Some scientists were quick to dismiss that possibility but were forced to reconsider their position when a second Martian meteorite (EETA7901) was found to contain the chemical signature of organic life.

The study of Mars meteorites was by no means the first attempt to determine whether life exists, or ever has existed, on Mars. This question has been argued and debated since the earliest days of the science of astronomy, and the debate has only intensified in recent years. NASA has launched a number of unmanned spacecraft to Mars and have openly stated that the search for life has been one of the main objectives of these missions, although they claim that to date, they have found no evidence of life on the surface or within the soil. However, many scientists, including some within the NASA community, have disputed these findings.

In 1976, the Viking lander missions conducted tests to look for evidence of life in the Martian soil, but Nasa stated quite publicly that the tests had proved negative, and that there were no signs of life in the soil samples tested. However, NASA scientist, Gilbert Levin, who played a key role in designing and developing these tests, said that the results were actually positive, and that they were erroneously misinterpreted. Dr Levin stated that a total of four positive tests, supported by five controls, were returned from two landers (Viking 1 and Viking 2) at a distance of 4,000 miles apart. Levin was adamant that the tests showed that the soil did contain micro-organisms and continued to stress this point until his death in 2001. Since that time, and on studying evidence from later Mars lander missions, there has been a shift in opinions at NASA and they have recently declared that they are quite open to the possibility of life on Mars, and even consider it quite likely.

So, why does it matter? Well, the point is that if there is, or ever has been, microbial life on Mars, which would prove that life can and does exist elsewhere in the universe and is not just confined to the Earth. If we take this a step further, it follows that as early life on Earth was once microbial and has evolved over time into the world as we know it today, there is no reason that, given the right set of circumstances, this process could not have taken place on other planets, too. The fact is that whilst scientists were debating the question of whether or not there is life on Mars today, others were advocating that Mars had actually harboured life sometime in the distant past, and in 1976 that notion was about to get a massive boost.

NASA's Mars mission, Viking 1 was launched in 1975 and in July 1976 it began orbiting Mars and sending pictures back to Earth. Several weeks into Viking 1's survey, NASA researcher Toby Owen was studying pictures of the Cydonia region of Mars when he saw what looked like a human face staring back at him. In frame 35A72 he saw what appeared to be a human face, with eyes, a nose and a mouth, on a mile-long raised plateau. The next day, NASA spokesman, Gerry Soffen held a press conference at which he showed the photo to a roomful of reporters and playfully stated, "Look what a trick of the light can do."

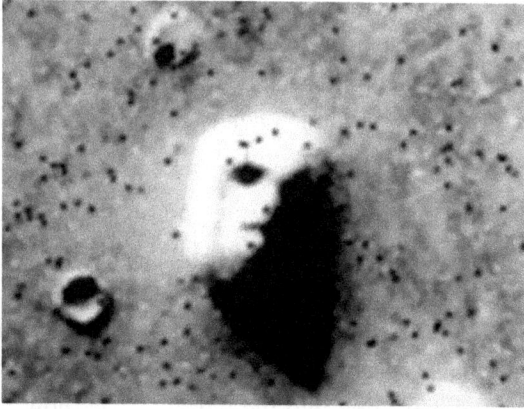

The face on Mars, as photographed by Viking 1 in 1976

He stated that although the image appeared to show a human face, it was actually a trick of the light and that a second picture taken a few hours later and in more favourable lighting conditions, showed no sign of any facial features. However, Soffen was unable to produce this second picture, as he said it had been misfiled. Although most of the gathered media accepted this explanation, some were sceptical. If the second picture had been taken so soon after the first, how could they have lost it?

Then in 1979, independent researcher, Vincent DiPietro set out to look for the missing photo and after trawling through thousands of images in the NASA archives, he found it. To his surprise, he discovered that the photo was taken not just a few hours, but 30 days after the picture that had been shown at the press conference. Furthermore, the orbital trajectory dictated that the spacecraft was nowhere near the Cydonia region at the time NASA had said the second photo had been taken.

Intrigued, DiPietro used state of the art imaging technology to compare the first photo with the second and was able to demonstrate that the image had not been a trick of the light but that it did, indeed, show a clearly defined representation of a human face. His conclusion was that this structure was either a bizarre act of nature, or it had been deliberately created by some form of intelligence.

As researchers began to look more closely at the photos taken by the Viking spacecraft, they began to see a number of other strange and intriguing structures on the surface of the red planet. Independent researchers have studied NASA's archive of Mars photos and have seen strange, five-sided pyramids, evidence of what look like ruined buildings and even a whole city, laid out in a geometric pattern.

The popular theory amongst ancient astronaut advocates is that at some time in the distant past, perhaps long before humans ever appeared on Earth, some unknown intelligent alien species either lived on, or perhaps visited Mars. They claim that the structures we see today, including the Cydonia face, are the remains of an ancient alien civilisation that once lived on Mars, but were forced to leave as the planet suffered some catastrophic event.

However, most scientists remain dismissive of these theories, and it is easy to see why. It would take quite a leap of faith to accept this hypothesis without firm evidence, and the general feeling amongst most planetary scientists is that these structures are almost certainly natural features. But there is always the possibility that they are not. If technologically advanced alien races are visiting Earth, and have been for millennia, why would they not visit, and even colonise, other planets? We know that Mars was once a far more hospitable place than it is now and could easily have supported intelligent life. If there were advanced alien races in the universe many millions of years ago, as many believe there are now, there is no reason why they would not want to colonise the red planet and extract its resources. The evidence we have for this, at present only circumstantial and the only way we can definitively answer the question, is to send people up there, or at least send a lander to the Cydonia region to investigate the area more closely.

The Moon

For as long as mankind has been on the Earth, we have gazed up in wonder at the Moon. It has been the one constant in an ever-changing world, but in all that time it has been shrouded in mystery. The Moon has always held a strange fascination for us here on Earth and for centuries we have tried to unravel its secrets. In recent years modern science has helped improve our understanding of the Moon but the conundrum has been that the more we learn, the more mysterious the Moon becomes.

The first baffling mystery is that the Moon should not be there at all. It is much larger in proportion to its host planet than any other moon in the solar system. Its size and mass are such that it is not really a moon at all, but rather a companion planet to Earth. The problem with that theory is that the Moon is nothing like the Earth, and so must have had a very different origin. That would suggest that the Moon may have been captured by the Earth at some time during the formation of the solar system, but that theory does not fit either, as the Moon is too big for that to happen. Science fiction writer and university professor, Isaac Asimov stated that, "The chances of such a capture having been effected, and the Moon then having taken up its nearly perfect circular orbit around the Earth are just too small to make such an eventuality credible."

Furthermore, it has now been established that the Moon is, in fact, hollow. During the Apollo missions seismic measuring equipment was placed on the Moon and when the now-redundant lunar module was deliberately crashed into the lunar surface, the Moon 'rang like a bell' for over an hour. This could only happen if the Moon was hollow. The late Carl Sagan stated that, "A natural satellite cannot be a hollow object. Therefore, the Moon may not be a natural satellite at all."

This exercise was repeated on Apollo 13 and on that occasion, the reverberation lasted almost three hours. For comparison, a similar impact on Earth would reverberate for just a few minutes.

Another curious fact is that according to many leading planetary scientists, our Moon has been placed precisely in the orbit in which it travels, and there is no way it could obtain its current orbit randomly. Ours is the only moon in the solar system that has a near-perfect orbit, which would suggest that something placed it there. And is it a coincidence that the Moon is just the right diameter, and is placed at just the right distance from the Earth, to perfectly cover the Sun during an eclipse?

All of this has led many, including some very eminent scientists and astronomers, to believe that the Moon is artificial and that it has been placed in its current orbit around the Earth quite deliberately by an alien intelligence sometime in the far-distant past. Many ancient civilisations tell tales of a time when the Moon was not there, and the Zulu people of Africa once held that the Moon is home to a race of reptilian aliens.

At this time, we have no way of knowing with any certainty whether the Moon is artificial or not, or how it obtained its current orbit around the Earth. But what we do know is that there are strange things happening on the Moon and have been for a very long time. The invention of the telescope in the late 16th century allowed scholars to study the Moon more closely and in greater detail than ever before, and many were surprised at what they saw.

In November 1668, colonial preacher, Cotton Mather was looking at the Moon through a telescope when he saw a mysterious light flying across the surface. In 1790, whilst observing an eclipse, astronomer Frederick Herschel saw what he described as, "Many bright and luminous points on the surface of the Moon, small and round in form."

Four years later, in 1794, Dr William Wilkins was observing the Moon from his home in Norwich when he saw a light, like a star, appear in the darkened area of the disc. He stated that the light was far-distant from the illuminated part of the moon and that the brightness increased, before gradually fading away. Moreover, in 1824 a Dutch selenographer reported seeing lights in the darkened area of the Moon, as did Thomas Elgar in 1867. Reports of this kind continued through the 19th century and even increased in frequency, but the advent of the space age and the formation of NASA in 1958 was a real game-changer in the study of anomalies on the Moon.

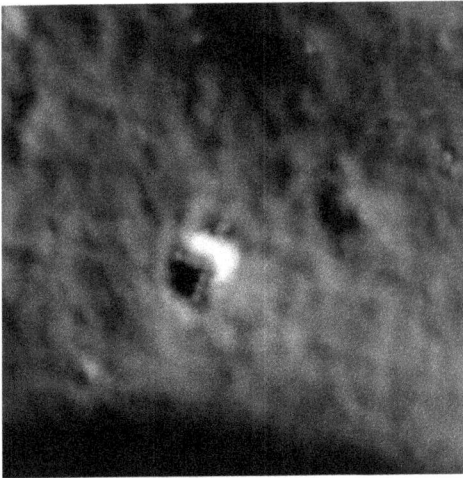

Strange formation of lights on the Lunar surface

Since its earliest days in operation, NASA has known about lights and other strange things on the Moon, which they call, Transient Lunar Phenomena (TLP), and they have compiled a document cataloguing these events, dating back to 1540. Furthermore, there is a firm belief in both scientific and ufology communities that NASA knows there are alien races on the Moon, and that the reason they went to the Moon in the first place was to investigate these claims. The theory is that NASA did find aliens on the Moon during the Apollo program and were told by them that we Earthlings were not welcome there. Could this be the real reason for NASA ending the Apollo program with Apollo 17 (missions were planned up to Apollo 20), and that they were warned off by hostile aliens?

Within a short time of the cessation of the Apollo moon landings, author, George E Leonard published his ground-breaking and controversial book, Somebody Else is on the Moon. In it he claims that on studying photos of the Moon's surface he was able to identify many large structures and anomalies that could only be artificial, supporting this claim with photographic evidence. He identified what he described as huge, ground-moving machines which he says is evidence that the Moon is being, or has been, mined by some unknown alien race. To support this theory, seismic measuring equipment placed on the Moon by the Apollo missions has recorded intense seismic activity around the western part of the Mare Nubium. NASA scientists have stated that these are the strongest signals they have seen on the Moon and that it is hard to see how they could be natural.

But if aliens are conducting mining activity on the Moon, what are they mining? One possibility is that they are mining Helium 3: a rare, stable isotope of helium. We now know that although rare on Earth, the Moon is rich in Helium 3, with the Earth having a total of around 15 tons, and the Moon having around 5 million tons. Helium 3 has the potential to be a powerful, non-polluting, non-radioactive source of fuel and it is thought that a tank the size of the one on the space shuttle could provide all of the energy needs of the US, cleanly and safely, for a whole year. Given these statistics, it is easy to see why the Moon might be of interest to an advanced alien race.

Leonard is not alone in his belief that the Moon is being mined by aliens. One former NASA contractor has stated publicly that although a well-kept secret at the time, it is common knowledge within NASA that the Apollo astronauts brought back pictures of alien structures on the Moon. Moreover, modern researchers such as Richard Hoagland and Mike Bara have been lobbying NASA for acknowledgement of this fact for years. Science writer, Ivan Sanderson has stated with some assertion that there are artificial structures all over the Moon and Nasa scientist, Dr Farouk El-Baz, who trained the Apollo astronauts on lunar geology, has also stated openly that there are strange things on the Moon.

Hoagland was able to obtain undoctored photographs of the Moon from outside of the NASA community that showed clear differences in photos of the same area, taken at different times. Craters seemed to change in shape and size and he identified dome-like structures in the floor of the crater, Tyco, which were not visible in later photos. Hoagland further claimed that video footage taken of the dark side of the Moon by Apollo 12 showed what he thought were large, glass domes, shattered and in a ruinous state.

When studied closely and in detail, many of the photographs in the NASA archive show features and objects on the Moon's surface that simply defy explanation. An image taken by Lunar Orbiter 3 in 1968 shows a view looking towards the horizon, where can be seen a mysterious tower, 1.5 miles high and seemingly made of glass. Photographs have also revealed another tower, this time, a staggering 20 miles high, located in the Mare Crisium region and known as the Shard. Furthermore, a photo taken by China's lunar rover, YUTU-2 shows a large, cube-shaped object sitting on top of a 7-mile-high tower. Other phenomena include triangular light formations that seem to hover above the Lunar surface and project light down onto it, as well as a labyrinth of parallel lines, dubbed, The City. Add to this, lights in the darkened areas of craters, large dish-like structures, bridges and even boulders that seem to move, leaving distinctive tracks that defy gravity by rolling uphill, and it is clear that there are things on the Moon that cannot easily be explained by conventional means.

Of course, it is entirely possible that these structures are, in fact, natural formations, but the evidence would suggest otherwise. It would appear that NASA and other government and scientific bodies know that there are artificial structures on the Moon, and possibly on Mars, but are trying to keep this knowledge from the public. But there have always been some in NASA and in other scientific organisations who feel strongly that the public should know the truth, and these brave stalwarts have helped to expose these secrets in a number of different ways.

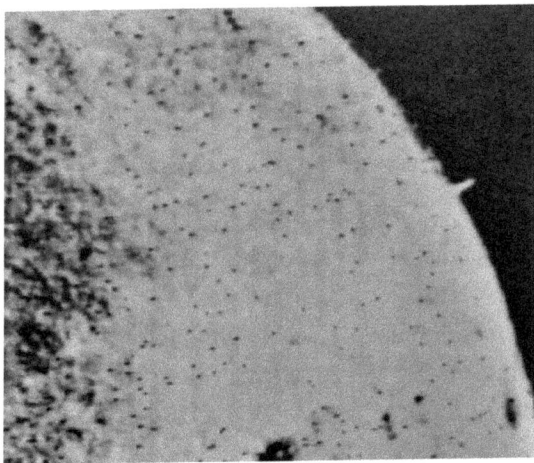

The so-called Lunar Shard, located in the Mare Crisium

In 1979, petroleum engineer Vito Saccheri, armed with a copy of George E Leonard's book, Somebody Else is on the Moon, began to pressure NASA to allow him access to their secret archive of unedited, high-resolution photos of the Lunar surface. He had it on good authority that many of the photos in the public domain had either been edited or were of poor quality, and he determined to gain access to the originals. He was stone-walled for some time but eventually, NASA relented and invited him to a secure facility to view the images in question. On entering the room he saw, laid out on tables, over 1,000 high resolution images of the Moon. He was not allowed to make copies or to make any notes, but he was given ample time to study the photos as closely as he wished, and he was shocked by what he saw. He said that his escort through the building was polite and friendly, but on entering the room the atmosphere changed and became almost solemn, "Because we knew what we were looking at, and we knew what it meant. The images were crystal clear and there were all kinds of structures, like bridges and towers and the like. All manufactured on a large scale."

Saccheri said that there was absolutely no doubt that the structures shown in the photos were artificial, as the images were very sharp and of very high quality. These structures are there, he said, and NASA has been hiding this from the public for decades.

Incredible as this may sound, these claims have been corroborated by many leading scientists and scholars over the past few decades. NASA scientist, Richard Whitcombe said that.

"Alien occupation of the Moon, either now or historically, is openly discussed and debated by NASA scientists. You will get quite a few qualified people to admit that there are strange things on the Moon. And in unguarded moments, some of them will admit their belief that these structures are of intelligent origin."

The Moon and Mars have proved fertile ground for the ancient astronaut theorists but their outlandish theories and beliefs are supported by serious scientific testimony. Some of world's leading scientists have put their careers and reputations on the line by stating quite openly that they believe there is, or has been, alien life on the Moon and on Mars, and perhaps in other parts of the Solar System, too.

Chapter 10: NASA and the Astronauts

Ever since its foundation in 1958, NASA pilots and astronauts have been seeing strange things in our skies and in space. One of the first to go public on this topic was NASA test pilot, Joseh Walker. In September 1962, Walker revealed that whilst flying the X-15 space plane at the very edge of the Earth's atmosphere two months earlier, he saw two disc-shaped objects flying alongside him at about the same altitude. He managed to film the objects but on studying the footage later, NASA scientists were unable to determine what they were. Furthermore, the following July, another X-15 pilot found himself in the middle of a formation of UFOs at a height of 314,000 feet, an altitude unachievable by any other known craft at that time.

NASA has always remained tight-lipped on the subject of UFOs, but there have been some insiders willing to break rank and speak out openly about their own experiences and about what they know. Gordon Cooper was one of the original seven NASA astronauts and was a firm believer in the reality of UFOs and that they were of alien origin, and that belief was based on his own personal experience. In 1953 Cooper was flying a fighter jet in the skies over Germany at its maximum height of 45,000 feet when he saw, ahead and above him, multiple disc-shaped objects flying much higher and faster than he was. He later stated.

"I could see they were not balloons or MIGs (Russian fighter jets) or even like any other aircraft I have seen before. They were metallic silver and were clearly disc shaped."

Cooper had another strange experience in 1963, when he was orbiting the Earth in his Faith 7 Mercury space capsule. He was talking to Ground Control on a secure frequency when strange-sounding voices broke in on the channel. The voices were recorded but when played back, they could not be matched to any known language on Earth. This strange phenomenon occurred again on some of the later missions.

NASA astronaut Gordon Cooper firmly believed that UFOs are real

Gordon Cooper made no secret of the fact that he believed in the reality of UFOs and he claimed that this view was shared by some of his fellow astronauts at NASA. He believed that UFOs could potentially pose a real threat to the security of the country and of the world and he felt that people should know the truth and that institutions, especially the military, should be more open and transparent in their dealings on the subject.

Cooper gave many interviews on the subject of UFOs, once stating that, "NASA and the government know very well that intelligent beings from other planets are visiting our world on a regular basis. They have an enormous amount of evidence but have kept quiet in order not to alarm people."

Project Mercury was followed by Project Gemini, a succession of two-man missions to test out rendezvous and docking procedures in Earth's orbit in preparation for Project Apollo, and these missions also encountered UFOs. In 1965, James McDivitt and Ed White were orbiting the Earth in their Gemini 4 space capsule when they saw a metallic object with antennae protruding from it. They were about to take pictures of the object when it came hurtling towards them at great speed, forcing them to take evasive action in order to avoid a collision. On another occasion, during the Gemini 12 mission, Buzz Aldrin photographed three glowing orbs that seemed to be tracking their spacecraft before disappearing at very high speed.

It is now widely accepted that NASA knows, and have done for many years, that UFOs are real and that they are of extra-terrestrial origin. It is claimed that they have given their astronauts code names to use when encountering UFOs in space, the best-known of these being, Santa Claus. Astronaut, Wally Shirra was thought to have been the first to use this term to refer to a UFO during his Earth-orbital mission in 1962. And there were others. On emerging from behind the Moon on the Apollo 8 Lunar-orbital mission in December 1968, astronaut Jim Lovell radioed back to mission control, "Please be informed, there is a Santa Claus."

NASA later claimed that as the communication took place on Christmas Eve, Lovell used this expression to express his relief that they had safely completed their orbit of the Moon. But many still believe that it was a genuine reference to a real UFO seen by the crew whilst on the dark side of the Moon and so, out of radio contact. Author and researcher, Maurice Chatelain has stated that all NASA missions were tracked and followed by UFOs. This view was shared by veteran astronaut, Scott Carpenter, who said.

"At no time while they were in space, were the astronauts alone. There was constant surveillance by UFOs."

A good example of this might be an incident that occurred during the Apollo 11 Moon-landing in 1969. As they were coasting towards the Moon, the crew noticed a bright object that appeared to be following them. They thought this might be the spent third stage of their Saturn V rocket booster but on checking with NASA, they were told that the booster was nowhere near them at that time, so could not have been the object they had been observing. They did not inform Mission Control of the reason for their query until after the mission and, on later investigation, NASA scientists were unable to establish what the object could have been and it remains unexplained to this day.

The Apollo program proved to be rich source of UFO sightings and of encounters with strange, unexplained phenomena. In May 1969, the crew of Apollo 10 reported seeing strange lights moving across the surface of the Moon, and Apollo 14 astronauts saw what they described as a large object with windows fly past their spacecraft whilst in lunar orbit. Staying with Apollo, photographs taken on Apollo 17 show a triangular shape in the black sky above the head of astronaut, Harrison Schmitt, as he walked on the Moon. It was claimed that this was a reflection, but as there was nothing to reflect, this did not satisfactorily explain the phenomenon. Moreover, as Apollo 17 flew over the Moon in lunar orbit, Schitt was heard to say. "Hey, I just saw a flash on the lunar surface. A bright flash, right out there near that crater."

In November 1969, the astronauts of Apollo 12 are said to have photographed a bright disc, and a similar object was observed and photographed by the crew of Apollo 13. Other pictures from the same mission showed a red cigar-shaped object seemingly tracking the spacecraft. There are many such reports associated with the Apollo program, but by far the most controversial and contested of these is an incident that is alleged to have occurred as the Apollo 11 astronauts stepped out onto the lunar surface.

In September 1979, after leaving the organisation, former NASA scientist Maurice Chatelain claimed that on stepping onto the Moon during the Apollo 11 mission, Neil Armstrong spotted two UFOs on the rim of a nearby crater and that they were filmed by Aldrin, who at that time was still in the lunar module. Radio communication with Earth was not live, but was on a delay so that, if necessary, Mission Control could edit the content before releasing it to the public. However, the live, real-time conversation was allegedly captured by amateur radio enthusiasts around the world, including Russia. According to the former Nasa Employee, Otto Binder, whilst on the lunar surface, a conversation took place between NASA and the crew of Apollo 11 in which Armstrong is alleged to have said.

"They are here. They are on the rim of a crater and they are watching us."

As might be expected, NASA has always denied that this exchange ever took place and the Apollo 11 astronauts have always dismissed the story as pure nonsense, stating that saw nothing unusual during their time on the Moon. It is a curious fact, however, that at a time when you would expect them to have been quite joyful and euphoric, all three Apollo 11 astronauts were noticeably quiet and reticent on returning to Earth and seemed reluctant to answer questions from the press. They seemed troubled and some even said, depressed, when one might have expected them to show contentment, or pride in a job well done. But we must remember that these were test-pilots, a breed not well-known for showing emotion or waxing lyrical. It is true that Neil Armstrong became very reclusive after the mission and shunned public attention, but he was a very private man, even before the mission. So, that does not necessarily mean that he saw something on the Moon that changed him.

With the early ending of the Apollo program in 1972, which some claim was a direct result of alien threats to stay away, NASA changed its focus from the Moon to Earth-orbit projects such as the Skylab space station and the space shuttle. But if they thought this would avert the attention of aliens and UFOs, they were quite wrong. If anything, UFO sightings and encounters increased in intensity throughout the Skylab and space shuttle programs, and on into the launch and operation of the international space station still in use today.

Skylab, NASA's first Earth-orbital space station, was a program designed to investigate and evaluate the effects of long-term exposure to the condition of weightlessness and to conduct various scientific experiments. It was manned by three 3-man crews through 1973 and 1974 and, during that time, its occupants reported numerous sightings of mysterious, unidentified objects seen in Earth's orbit. For example, in 1973, astronauts Jack Lousma, Owen Garriot and Al Bean were orbiting the Earth aboard Skylab when they saw a rotating red object out of the window. They spent ten minutes photographing the object and on studying the photos later, NASA said it was another spacecraft. However, there were no other known space vehicles on the vicinity at that time.

As the space shuttle program followed Skylab, the mysterious sightings continued and were becoming more extensively reported and difficult to ignore. For example, in September 1991 a video released by Nasa and taken aboard the space shuttle in Earth's orbit, showed a number of strange objects flying through space. One of these objects can be seen firing a beam of light at the other, which then shoots out of the frame at fantastic speed. NASA claimed that the objects were particles from the shuttle's thrusters but many, including leading scientist Dr Jack Casher, rejected this claim and stated that the objects were clearly under intelligent control and so, were not explainable. On another occasion during shuttle mission STS-80 in 1996, two disc shaped objects were seen close to the shuttle orbiter. Again, NASA tried to offer and explanation and again, this was rebuffed by the experts. UFO sightings by space shuttle crews became so frequent and commonplace that NASA ceased to broadcast live transmissions from space to avoid having to explain unexplainable.

Chapter 11: Government Cover-Up?

Despite decades of denial and feigned indifference, the United States and other governments have always taken a far keener interest in UFOs than they have ever let on. As UFO sightings increased in the years following WWII, governments around the world were at a loss to know how to handle this new and mysterious phenomenon. Winston Churchill, Prime Minister of Great Britain, and the then US President, Harry Truman both asked their intelligence agencies to investigate UFO sightings and report back to them on their findings. Their fear was that there were strange craft operating in their air space and they had no idea what they were or where they came from, and governments around the world were unsure how to respond to this potential threat.

The United States had emerged from WWII a global superpower, and at this time was engaged in a dangerous cold war with the Soviet Union. There were fears that these craft might be of Russian origin, built using technology gained from captured Nazi scientists. On the other hand, there was always the real possibility that they might be extra-terrestrial in origin, as some were already beginning to advocate. Either way, if they did pose a hostile threat, their technological superiority was such that the armed forces of the time would be powerless against them.

The decision to keep the UFO phenomenon a closely guarded secret began with the now famous Roswell case in 1947. Here, the US Air Force issued a press release stating that a UFO had crashed in the desert and that wreckage, and even alien bodies, had been recovered from the scene. But almost immediately after this story appeared in the newspapers, it was swiftly retracted and a revised statement declared that what had been recovered was wreckage from a downed weather balloon.

This denial was meant to create the impression that the US military saw nothing unusual in this event and that they had no real interest in UFOs, but nothing could have been further from the truth. Following the Roswell incident, President Harry Truman instructed the Secretary of Defence, James Forrestal, to set up a secret committee to investigate this, and other similar cases. The result of this directive was the formation of a project code-named Majestic 12, which was to be led by leading scientist, Vannevar Bush. The fact is that, despite his casual indifference to the UFO threat, President Truman was worried, stating that, "I can assure you that, given that these flying saucers exist, they are not constructed by any power on Earth."

Majestic 12 was replaced in 1948 by Project Sign, which in turn was replaced by Project Grudge in 1949 and again by Project Blue Book in 1952; all of them top secret projects, and all of them tasked with finding out all they could about UFOs.

Many leading scientists were invited to take part in these investigations, most of them sceptics, chosen by the government for their conviction that UFOs simply did not exist. But on examining hundreds of cases and interviewing thousands of witnesses, some of them began to have a change of heart. One such dissenter was Dr J Allen Hynek, professor of astronomy at the Ohio State University. On joining Project Sign, Hynek was one of the most ardent sceptics in the group and was of the opinion that all UFO sightings could be explained by scientific means. However, interviews with witnesses, including both military and civilian pilots, caused him to rethink his position and he became convinced that UFOs were very real and that they were of extra-terrestrial origin.

Launched in 1952, Project Blue Book was terminated in 1969 when physicist Edward Condon published his government-sponsored report on the UFO phenomenon. On examining cases from Project Blue Book, Condon concluded that there was no hard evidence to prove that UFOs were real and that further study was unlikely to yield any major scientific discoveries. But the Condon report had its critics and there were many who believed that the purpose of the report was to deceive the public into thinking that the government had no further interest in UFOs, whilst continuing to study them in secret. One such sceptic was renowned physicist Dr James McDonald.

Having witnessed a UFO sighting himself in 1954, Dr McDonald firmly believed that UFOs were real and that the US government were of the same mind. He campaigned hard for more transparency and for more in-depth research, and he suggested that the real objective of Project Blue Book was to debunk and ridicule witness accounts. In 1967, following years of intensive research, Dr McDonald said,

"There is no sensible alternative to the utterly shocking hypothesis that UFOs are extra-terrestrial probes."

McDonald campaigned at the highest levels in government and in 1968 he stated before Congress that, "UFOs are entirely real, and we do not know what they are because we have laughed them out of court."

Dr McDonald's tenacity gained him many enemies and in 1971 he died in what some thought to be suspicious circumstances. Just days before his death he had told fellow physicist, Dr Robert Wood, "I am very close and will soon have all the answers. I just need more time."

Dr Woods said that McDonald showed no signs of depression at the time and in fact, seemed eager to continue his work. McDonald's body was found in the desert with a bullet wound to the head. A .38 revolver was found by his side, so the death was declared to be a suicide, but many suspected foul play.

So, who is keeping what secret, and from whom? It would appear that there are powerful groups within government and intelligence agencies around the world that are operating at such extreme levels of secrecy, they are not even informing their own leaders of their activities. This policy of secrecy and exclusion began during the closing months of the Eisenhower administration, prompting him, on leaving office, to make the following statement.

"We must guard against the acquisition of unwarranted influence, whether sought or unsought, by the military-industrial complex."

Eisenhower was well-qualified to make such a statement. He is reported to have witness a UFO sighting himself before becoming president, and he is known to have believed in life on other planets. Furthermore, there are some unexplained gaps in Eisenhower's diary for February 1954 which have been the source of much controversy and debate. The official explanation was that the president was on holiday in Palm Springs, but witnesses have come forward with the incredible claim that during this missing time, the president was meeting with the Nordic-type aliens at Holloman Air Force base, New Mexico. The theory behind this outlandish claim was that the US government had made an agreement with the aliens that they would turn a blind eye to their sinister activities here on Earth, in return for their superior technology. Whatever the truth, Eisenhower would be the last US president to be completely privy to the workings of the military and intelligence communities on the subject of UFOs as the 'need to know' culture took hold.

One of the first US presidents to feel the brunt of this cold-shoulder tactic from their first days in office was Richard Nixon. He believed that UFOs were real and that they were of extra-terrestrial origin and he wanted to claim a place in history by being the first world leader to reveal the truth about UFOs, but he was denied that privilege, as was Jimmy Carter a few years later. Carter was fascinated by the subject of UFOs, having witnessed one himself in 1969, and this experience prompted him to declare, "One thing is for sure,

I'll never again make fun of people who say that they have seen a UFO."

Like millions of other people around the world, Carter was keen to learn all he could about UFOs and he felt strongly that the public should know the truth, whatever that may be. When campaigning for the US presidency in 1976, Carter stated that, "If I become president, I will make every piece of information this country has about UFO sightings available to the public and to the scientists."
However, on gaining office, Carter soon learned that he would be in no position to honour that promise, and he did not make it again.

Bill Clinton had a similar experience on his accession to the White House. In this case, Clinton appointed close friend Webb Hubble to the position of Associate Attorney General, saying, "Webb, if I put you over at justice, I want to you to find the answers to two questions for me; who killed JFK, and are UFOs real?" Hubble later said that he investigated both topics and was not satisfied with the answers he was getting. Furthermore, when questioned on the subject by a schoolboy whilst on a trip to Ireland, Clinton said, "Ryan, if the United States Air Force did recover alien bodies from Roswell, they didn't tell me about it either. And I want to know."

Of all the US presidents since WWII, Ronald Reagan has been the most outspoken on the subject of UFOs. The Reagan administration governed at the height of the cold war with Russia and Reagan worked tirelessly to improve relations between the United States and the Soviet Union, and this may have been partly due to what he saw as a potentially universal alien threat. Whilst walking around Lake Geneva with Mikhail Gorbachev during a summit in 1985, Reagan is reported to have asked Gorbachev, "What would you do if the United States were suddenly attacked by someone from outer space, would you help us?"

To which Gorbachev replied, "No doubt about it. We too have similar concerns."

There are limits to what any world leader can or will say on the subject of UFOs, but Ronald Reagan seemed eager to say more than most. In a speech to the United Nations in 1987, Reagan urged members to imagine how quickly humanity would come together if it were to be confronted by an extra-terrestrial civilisation, saying, "What if all of us in the world were threatened by a power from outer space or from another planet? We would suddenly find out that we didn't really have any differences at all."

He later added, "Perhaps we need some outside, universal threat to make us recognise this common bond."

Bold words for a sitting US president sworn to secrecy!

Chapter 12: End Game?

In December 2017, the New York Times published two videos leaked to them by Christopher Mellon, former Deputy Assistant Secretary of Defence for Intelligence at the Pentagon. The footage was taken from the gun cameras on US Navy fighter jets and appeared to show the aircraft chasing a group of mysterious, tic-tac shaped UFOs which demonstrated incredible flying abilities, far beyond those of any known conventional aircraft. The footage very quickly went viral on the internet and there were increasing calls, even from within government, for the Pentagon to reveal where and when the videos had been taken. Unable to withstand the ever-growing tide of public pressure, the Pentagon finally agreed to release the videos in April 2021, along with an official statement that the footage was real, and that they were still unable to identify the objects shown in the films. Following this, some of those involved in the incident began to speak out and tell the world exactly what had happened, and the full story began to emerge.

That story was that in November 2004, the aircraft carrier USS Nimitz was one of a fleet of vessels on a routine exercise off the coast of California when their control ship, the USS Princeton began to see numerous anomalies on their SPY1 radar screen. Senior operations specialist, Kevin Day revealed that over a period of several days he saw multiple unidentified objects, as many as ten at any one time, on his radar screen. He judged that they were clearly not any conventional aircraft, as they displayed flying abilities which were simply unachievable by any known aircraft at that time. They could drop from 28,000 feet to just 50 feet in less than one second, equating to a velocity of around 24,000mph and a G-force of around 1,350Gs. Furthermore, these objects could change direction instantaneously and although travelling through the atmosphere at phenomenal speeds, they were silent and produced no sonic boom.

The objects appeared on the radar again on November 14th and this time, the controllers on board Princeton were able to vector two F18 Super Hornet jet fighters from the aircraft carrier USS Nimitz to intercept. At first the pilots could see nothing, either visually or on their on-board radar, but as they drew closer to their vector point, they saw a strange disturbance on the surface of the ocean about two miles away. On reaching the point of the disturbance they saw a white, tic-tac shaped object, about 47 feet long, which appeared to be moving erratically, just like the object seen on Princeton's radar.

The object climbed up level with the jets and flew alongside them, mirroring their manoeuvres and banking left and right with them, before flying off into the distance at incredible speed. The pilots had seen nothing like this before and were certain that the object was under intelligent control. Ops specialist, Kevin Day said that the pilots were in constant communication with him by radio and that as they described what was happening, he could hear the fear and shock in their voices. The jets followed the object and managed to capture video footage of it flying low over the water, and at times, beneath the surface, creating a disturbance in the sea in its wake. A second pair of Super Hornets was launched later in the day and they, too, had a similar experience that they were able to capture on film. These videos were those released by the Pentagon in 2021.

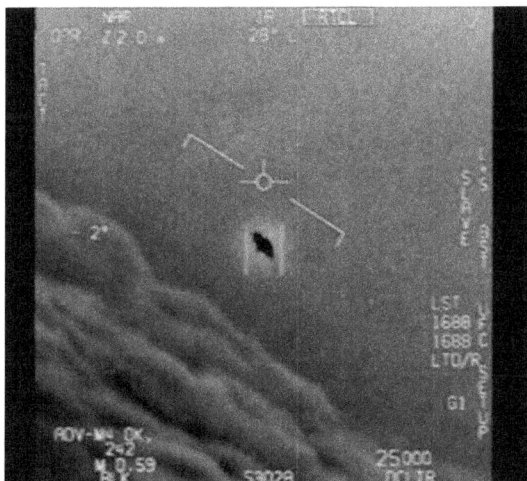

UFO captured on gun-camera film by aircraft from USS Nimitz

On returning to the carrier, the pilots prepared for their routine de-briefing when events began to take a more sinister turn. They were escorted to a secure room aboard the USS Nimitz, where they were ordered to sign a non-disclosure secrecy agreement and told never to discuss the incident with anyone. Moreover, when Petty Officer Patrick Hughes removed the hard drives containing all the flight data from the Super Hornets, in line with standard practice, they were ceased and confiscated by two men he had never seen before and taken away by helicopter. Similarly, the radar data from Princeton was also removed and taken away, and all on-board files deleted.

The USS Nimitz incident was a real game-changer in the field of ufology and was the watershed for all that was to follow. In response to growing pressure from the media and the UFO research community, officials at the Pentagon were forced to admit that, not only were they secretly studying UFOs, but also that they had strong evidence so show that they do exist and that they had no idea what they were or where they came from.

In light of recent revelations and official government statements on the subject, we now have a far better understanding of the technological capabilities of UFOs, and it is this advanced level of sophistication that makes it difficult to conclude that they are of Earth's origin. A typical UFO is able to navigate any distance in any environment and can move as freely in the water as in the air. They have no wings or engines in the conventional sense and would appear to have the ability to defy the effects of the Earth's gravitational pull. Moreover, they leave no visible or audible trace. They leave no vapour trails and although travelling at phenomenal speeds, they generate no sonic booms.

UFOs have been tracked travelling at speeds up to 38,000mph, or Mach 17, which would be impossible for any conventional aircraft. Any known aircraft flying at anything beyond Mach 3, would simply disintegrate, killing the pilot and completely destroying the aircraft. Furthermore, it would be impossible to fly through the air at these speeds without generating multiple sonic booms, as mentioned above, but UFOs can travel at these, and even greater speeds, in complete silence. Another noteworthy attribute of UFOs is their incredible rate of acceleration. Conventional aircraft accelerate and decelerate gradually, and at a controlled rate. What is more, they turn in a similar way, banking into gradual turns that require miles of airspace to complete. Conversely, UFOs are capable of sudden, rapid acceleration, often starting at one point then accelerating at great speed to another point many miles away in seconds, before coming to a sudden and dead stop, or changing direction again. There is not an aircraft in the world that can do that.

Through studying the Nimitz incident, and other similar cases, we have been able to determine some of the technological capabilities of UFOs, and what we have seen only strengthens the argument that they are not of Earth's origin. This hypothesis has been gaining momentum in recent years and we are now seeing declarations of support for the idea from people who, just a short time ago, were hard-line sceptics. Many in the scientific community and in government are now far more open to the possibility that UFOs are real and that they may be of extra-terrestrial origin. Former US Senator, Harry Reid recently declared, "Are they a threat? I don't know. But what I do know is that we can't just turn our heads and say they don't exist, because they do exist."

Assuming that, based on the overwhelming body of evidence, UFOs are real and that they are from elsewhere in the universe, do the governments of the world know this and are they ever going to tell us? In recent years we have seen a gradual but significant shift toward a policy of openness and transparency on the subject of UFOs, and that has helped to encourage more people to speak out. Paul Hynek, UFO researcher and son of Project Blue Book's Dr J Allen Hynek, stated that.

"There has recently been a shift in the communication policies of the US government and what we are now seeing is a gradual process towards a grand disclosure."

It has long been argued that the reason for the government's secrecy on the subject of UFOs was that they feared mass panic and social meltdown if the public were told that we are not alone in the universe and that we are, in fact, being visited by alien races. This fear may have been well-founded in the years following WWII, but evidence would suggest that this would no longer be the case. Despite decades of denial, debunking and ridicule, recent surveys have shown that around 65% of Americans believe in extra-terrestrials, and that number is increasing all the time. This increase was further enhanced in 2021 by the release of the ODNI (Office of the Director of National Intelligence) report on Unidentified Aerial Phenomena, in which it was stated that of the 144 cases examined, 143 remained unidentified.

If aliens do exist, and if the government knows this to be true, then it is time we were told the truth. The public are ready for such a disclosure and it may, in fact, already be happening. Such a disclosure would likely be a gradual process, with snippets of information being trickled out at a controlled rate. In 2018 Luis Elizondo, former Director of AATIP (Advanced Aerospace Threat Identification Program), made the following statement when addressing a meeting of MUFON (Mutual UFO Network), "Disclosure is not an event, it is a process. And that process has already begun."

There are so many questions to consider when trying to solve the UFO riddle, many of which still remain unanswered. People may justifiably ask, why do aliens abduct very ordinary people and not scientists or world leaders? How is it that a technologically advanced alien spacecraft can travel vast distances across interstellar space, only to crash in the desert or on a mountain in Wales? And if aliens are here, why have they not shown themselves? These are all good and valid questions, and we may never know the answers. The UFO phenomenon is like a huge jigsaw puzzle, and we cannot solve this puzzle at this time because we do not have all the pieces.

If UFOs are real and if aliens are amongst us, then we should know the truth. On the other hand, if there is nothing to tell, why not just come out and say so? The world is ready to know the truth about UFOs, one way or another, and until that truth is revealed there will always be doubt and speculation.

www.ingramcontent.com/pod-product-compliance
ning Source LLC
rsburg PA
030210326
00009B/1067